Exercises in Synthetic
Organic Chemistry

Exercises in Synthetic Organic Chemistry

CHIARA GHIRON and RUSSELL J. THOMAS

Department of Medicinal Chemistry,
Glaxo Wellcome S.p.A.
Verona,
Italy

OXFORD NEW YORK TOKYO
OXFORD UNIVERSITY PRESS
1997

Oxford University Press, Great Clarendon Street, Oxford OX2 6DP
Oxford New York
Athens Auckland Bangkok Bombay Buenos Aires
Calcutta Cape Town Dar es Salaam Delhi Florence Hong Kong
Istanbul Karachi Kuala Lumpur Madras Madrid Melbourne
Mexico City Nairobi Paris Singapore Taipei Tokyo Toronto
and associated companies in
Berlin Ibadan

Oxford is a trade mark of Oxford University Press

Published in the United States
by Oxford University Press Inc., New York

A catalogue record for this book is available from the British Library

Library of Congress Cataloging in Publication Data
(Data available)

ISBN 0 19 855944 5 (Hbk)
ISBN 0 19 855943 7 (Pbk)

Typeset by the authors
Printed in Great Britain by
Butler & Tanner Ltd,
Frome, Somerset

This book is dedicated to our families

Preface

The advent of ever-more sophisticated methods of information retrieval is revolutionising the way chemists work. The possibility of accessing a database which, in a matter of seconds, is capable of providing hundreds of methods of carrying out a synthetic transformation means that the time in which a synthetic strategy can be planned is reduced enormously.

A subtle, but no less profound effect of this completely new approach is in the way chemists handle the 'vocabulary' of their profession, a knowledge of possible chemical transformations. It could be said that it will become less important to memorise lists of synthetic methods, but this creates a problem. Computer reaction databases are only as good as the questions we ask them, and without a sound knowledge of what is chemically feasible, we cannot construct a query to obtain the exact reaction conditions we need.

Undoubtedly one of the most powerful tools for the design of a synthesis is the elegant concept of retrosynthesis. By looking at the target structure, and having the knowledge of how each of the functional groups, and some of the carbon–carbon bonds, present in the molecule could be introduced, it is possible to dissect the target, taking it back to potential starting materials. This concept again relies on the chemist having a large enough vocabulary of transformations to hand.

One of the best ways of increasing a person's knowledge of the chemical transformations available is to spend time analysing published syntheses. By following a molecule through the various transformations to the final product it is possible to observe, in complex real-life situations, the application of synthetic methods. An even more effective approach is to study an article in the form of a synthetic exercise, either for informal discussion in a group, or private study. This has the benefit of encouraging the chemist or student to reflect for a while on possible mechanisms, reaction conditions, and the stereochemical outcome without having the answer immediately to hand.

An additional benefit, over simply studying an article from the literature, is that in some way it removes the choice of topic. This may be important bearing in mind the natural tendency of many of us to study areas that we already know something about, thus reducing the learning curve. This book has therefore deliberately chosen examples from a wide range of synthetic targets.

The purpose of this book is to provide chemists with a collection of exercises constructed from the recent literature. The exercises are designed to try and provide people at various levels with synthetic challenges, from final year undergraduates to graduate students and more experienced post-doctoral chemists. Thus in a group consisting of final year undergraduate students, post-graduate research students, and more experienced post-doctorate and academic staff members, the less experienced members can learn about the more fundamental organic transformations, protecting group strategies and stereochemical considerations, while the more senior chemists have the possibility to analyse the more detailed mechanistic aspects, while having the opportunity to revise and discuss the basic concepts.

We hope that this text proves useful both in academic and industrial chemistry departments, and may provide the basis for productive group discussions of synthetic problems. We shall always be pleased to receive comments and suggestions from readers as to how we can improve on the concept for future volumes.

Verona   C. G.
October 1996   R. J. T.

The Authors

Chiara Ghiron was born in Genova, Italy in 1965. Having read chemistry at The University of Genova between 1985 and 1990 she joined Glaxo Wellcome in Verona, where she is currently a member of the Medicinal Chemistry Department.

Russell J. Thomas was born in Swansea, Wales in 1966, and read chemistry at the University of Kent at Canterbury between 1984 and 1987. Having completed his Ph.D. with Prof. Stan Roberts at the University of Exeter in 1990, he moved to Verona in Italy to work for Glaxo Wellcome in the Medicinal Chemistry Department.

Acknowledgements

The authors would like to thank Phil Cox, Sylvie Gehanne, Fabrizio Micheli, and Maria Elvira Tranquillini for help in proof-reading the exercises. Thanks also to Daniele Donati, Tino Rossi, and Melissa Levitt for encouragement and helpful suggestions, and to the staff at OUP for their support of the project at its various stages. The authors would also like to acknowledge Glaxo Wellcome S.p.A. for granting permission to undertake the writing of this book.

Contents

Contents

Common Abbreviations Used in the Text

AD	Asymmetric dihydroxylation
AE	Asymmetric epoxidation
AIBN	α,α'-Azo*iso*butyronitrile
Aldrithiol®	2,2'-Dipyridyldisulphide
All	Allyl
Alloc	Allyloxycarbonyl
9-BBN	9-Borabicyclononane
BMS	Borane-methylsulphide complex
BINAP	2,2'-Bis(diphenylphosphino)-1,1'-binaphthyl
BINAPO	Phosphinous acid, diphenyl-[1,1'-binaphthalene]-2-2'-diyl ester
Boc	*tert*-Butoxycarbonyl
BOM	Benzyloxymethyl
BOP-Cl	Bis(2-oxo-3-oxazolidinyl)phosphinic chloride
Bn	Benzyl
BTAF	Benzyltrimethylammonium fluoride
Bz	Benzoyl
BHT	*tert*-Butylhydroxytoluene
CAN	Ceric ammonium nitrate
CBS	Corey-Bashki-Shibat
Cbz	Benzyloxycarbonyl
Cp	Cyclopentadienyl
*m*CPBA	*meta*-Chloroperoxybenzoic acid
CSA	Camphorsulphonic acid
DAST	Diethylaminosulphur trifluoride
dba	Dibenzylideneacetone
DBU	1,8-Diazabicyclo[5.4.0]undec-7-ene
DCC	1,3-Dicyclohexylcarbodiimide
DDQ	2,3-Dichloro-5,6-dicyano-1,4-benzoquinone
DEAD	Diethyl azodicarboxylate
DHP	Dihydropyran
DHQD	Dihydroquinidine
DIBAL	Diisobutylaluminium hydride
DIBAL-H	Diisobutylaluminium hydride
DIC	Diisopropylcarbodiimide
DMAP	4-Dimethylaminopyridine
DME	Dimethoxyethane
DMF	*N,N*-Dimethylformamide
DMP	Dess-Martin periodinane
DMPM	3,4-Dimethoxybenzyl
DMS	Dimethylsulphide
DMSO	Dimethylsulphoxide
DPPA	Diphenylphosphoryl azide

EE	1-Ethoxyethyl
FDPP	Pentafluorophenyl diphenylphosphinate
Fmoc	9-Fluorenylmethoxycarbonyl
HMPA	Hexamethylphosphoramide
HMPT	Hexamethylphosphorous triamide
Ipc	Isopinocampheyl
KDA	Potassium diisopropylamide
KHMDS	Potassium bis(trimethylsilyl)amide (KN(TMS)$_2$)
K-Selectride®	Potassium tri-*sec*-butylborohydride
LDA	Lithium diisopropylamide
LDEA	Lithium diethylamide
LiHMDS	Lithium bis(trimethylsilyl)amide (LiN(TMS)$_2$)
MEM	2-Methoxyethoxymethyl
MOM	Methoxymethyl
MPM	*p*-Methoxybenzyl
MS	Molecular sieves
Ms	Methanesulphonyl
MW	Microwave
NaHMDS	Sodium bis(trimethylsilyl)amide (NaN(TMS)$_2$)
NBS	*N*-Bromosuccinimide
NMM	*N*-Methylmorpholine
NMO	*N*-Methylmorpholine-*N*-oxide
N-PSP	*N*-Phenylselenophthalimide
Ns	*p*-Nitrophenylsulphonyl
PDC	Pyridinium dichlorochromate
PCC	Pyridinium chlorochromate
Ph	Phenyl
Piv	Pivaloyl
PMB	*p*-Methoxybenzyl
PMP	*p*-Methoxyphenyl
PPA	Polyphosphoric acid
PPL	Porcine pancreatic lipase
PPTS	Pyridinium *p*-toluenesulphonate
PTSA	*p*-Toluenesulphonic acid
Pv	Pivaloyl
Py	Pyridine
PyBroP	Bromotripyrrolidinophosphonium hexafluorophosphate
SEM	[2-(Trimethylsilyl)methyl
SEMCl	[2-(Trimethylsilyl)ethoxy]methyl chloride
Red-Al®	Sodium bis(2-methoxyethoxy)aluminium hydride
TBAF	Tetrabutylammonium fluoride
TBDMS	*tert*-Butyldimethylsilyl
TBDPS	*tert*-Butyldiphenylsilyl
TBS	*tert*-Butyldimethylsilyl
TCDI	Thiocarbonyl diimidazole

Common Abbreviations Used in the text

Tf	Trifluoromethanesulphonyl
TFA	Trifluoroacetic acid
TFAA	Trifluoroacetic anhydride
THP	Tetrahydropyranyl
TIPS	Triisopropylsilyl
TMG	Tetramethylguanidine
TMS	Trimethylsilyl
TMSCl	Trimethylsilyl chloride
TMSOTf	Trimethylsilyl trifluoromethanesulphonate
Tol	*p*-Toluyl
o-Tolyl	*o*-Toluyl
TPAP	Tetra-*n*-propylammonium perruthenate
Tr	Trityl
Troc	2,2,2-Trichloroethoxycarbonyl
Ts	*p*-Toluenesulphonyl
p-TsOH	*p*-Toluenesulphonic acid
Vitride®	Sodium bis(2-methoxyethoxy)aluminium hydride
Z	Benzyloxycarbonyl

Introduction

The purpose of the book

The exercises in this book are intended to provide challenges for people with various levels of experience. A final year undergraduate student should obviously not be expected to tackle a problem without the aid of his or her favourite textbooks, and will still undoubtedly have difficulty with the more advanced problems. The advanced aspects of an exercise are intended for a more experienced chemist to analyse and discuss in detail. A student will hopefully find that in studying the exercises, while at first it will be difficult to complete even half of the questions unaided, with time both the size of his or her 'vocabulary' of reactions and the time necessary to study an exercise will change dramatically.

How an exercise is constructed

The problems are taken from recent publications either of total syntheses of natural products, or the syntheses of related systems. The answers to the various questions are not provided in this book, although they can easily be obtained from the original articles. This was done not only to reduce the size of the text, but also to allow study and discussion of a problem in at least a formal situation of not knowing the answer. Seeing a reaction in which an olefin is transformed into a 1,2-diol with osmium tetroxide is a useful way of learning chemistry, but not nearly as effective has having time to think of what the reagent or product could be *before* seeing the solution.

The article, or articles from which the exercise was taken is cited immediately below the title. In the schemes, some of the structures or reaction conditions (in bold letters) are missing. At the end of the scheme there are some additional discussion points, where additional questions regarding mechanism, choice of reagents, stereochemistry, stereoselectivity etc. are listed. Below the discussion points there are usually some additional articles and reviews for further reading on key topics covered in the exercise. These references are usually not cited in the original paper. In order to construct a coherent exercise, it was necessary to change the molecule and reaction numbering used in the original publication.

Although it is a very subjective choice, the exercises are ordered in an approximately ascending order of difficulty. While obviously there can be no absolute guarantees, a final year undergraduate or more junior postgraduate student is advised to start at the front and work through the book from there.

In many cases the target molecules have been synthesised by several research groups around the world with quite differing approaches. Our choice of which article to use in this book should not be considered as a measure of which approach is more valid or innovative, the choice was made simply on the basis of how the syntheses offered themes to be explored in this book. Similarly, we do not intend that articles chosen for exercises appearing at the start of the book should be considered in any way less innovative than those chosen for more difficult exercises towards the back. Clearly the difficulty of an exercise largely depends on the amount of information provided or omitted.

How to approach an exercise

As this book is intended for chemists with a wide range of experience, there is no single best way of approaching an exercise. However, possibly the most effective way for everyone is to first try and complete as much as possible unaided before going back over the more problematic parts with the help of additional textbooks and discussion with colleagues. Having completed these first two passes, a final check of both the answered and unanswered problems can be made with the help of the original article. It is not too surprising how obvious most of the reactions become with the aid of the original text to hand when studying an exercise! An additional benefit of not consulting the original paper until the end is that it gives the possibility of proposing alternative approaches to those used, which can then be discussed with colleagues.

In cases where an unknown structure cannot be deduced from the reagents used to form it, a useful alternative is to look at the next structure in the synthetic sequence and go backwards, using a form of retrosynthetic analysis based on the reagents necessary to arrive at the known structure.

An important concept for the less experienced chemists is that we would be very surprised if someone (other than the authors of the original journal article!) were able to complete all of the questions exactly. With time and experience, the number of unanswered questions will decrease notably, although hopefully it will never arrive at zero.

1. Total Synthesis of (−)-Ovatolide

A. Delgado and J. Clardy, *J. Org. Chem.*, **1993**, *58*, 2862.

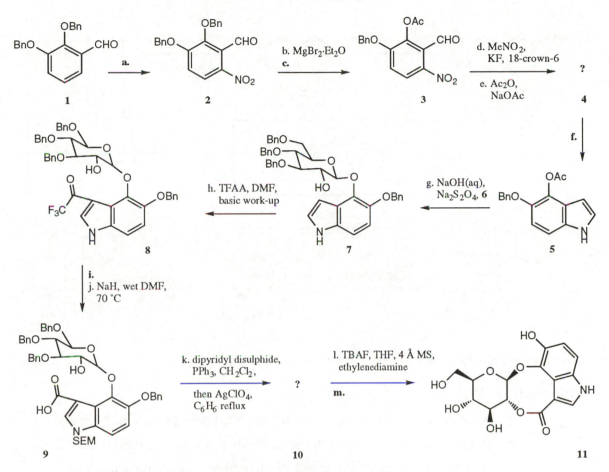

Discussion Points

- What other isomer is formed in the nitration step **a**?
- Rationalise the regioselectivity of the mono-debenzylation reaction carried out in step **b**.
- Suggest a structure for compound **6** and give reasons for carrying out the reaction in a reductive environment.
- What is the mechanism of the hydrolysis of compound **8**?
- Give an explanation for the use of ethylenediamine in step **l**.

Further Reading

- For a review on diastereoselective nitroaldol reaction, see: D. Seebach, A. K. Beck, T. Mukhopadhyay and E. Thomas, *Helv. Chim. Acta*, **1982**, *65*, 1101. For a modification of the nitroaldol reaction using alumina, see: G. Rosini, E. Marotta, P. Righi and J. P. Seerden, *J. Org. Chem.*, **1991**, *56*, 6258.

2. Total Synthesis of RS-15385

J. C. Rohloff, N. H. Dyson, J. O. Gardner, T. V. Alfredson, M. L. Sparacino and J. Robinson III,

J. Org. Chem., **1993**, *58*, 1935.

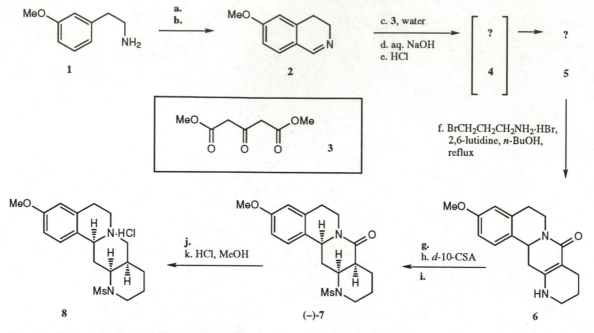

Discussion Points

- What is the mechanism of the Bischler–Napieralski reaction of step **b**?
- Suggest a reason why less hindered bases could not be used in step **f**.
- Propose structures for the minor diastereomers obtained during the hydrogenation step **g**.
- Suggest a derivatising agent for the HPLC analysis of the optical purity of the 10-camphorsulphonic acid salt obtained in step **h**.

Further Reading

- For reviews on isoquinoline alkaloids synthesis, see: M. D. Rozwadowska, *Heterocycles*, **1994**, *39*, 903; E. D. Cox and J. M. Cook, *Chem. Rev.*, **1995**, *95*, 1797.
- For a review on chiral derivatising agents, see: Y. Zhou, P. Luan, L. Liu and Z.-P. Sun, *J. Chromatogr. B: Biomed. Appl.*, **1994**, *659*, 109.

3. Total Synthesis of Islandic Acid I Methyl Ester

T. Shimizu, S. Hiranuma, T. Watanabe and M. Kirihara, *Heterocycles*, **1994**, *38*, 243.

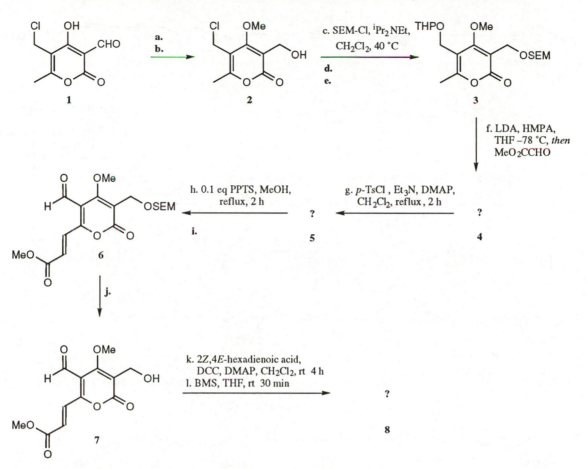

Discussion Points

- The sequence from **2** to **3** introduces a tetrahydropyranyl (THP) group. What is the mechanism of this step?

- Following step **d** a quantity of compound **9** was isolable. Suggest a mechanism for its formation.

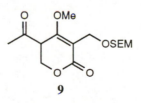

9

- Give two alternative methods of removing a SEM group as in step **j**.

4. Total Synthesis of (+)-Patulolide C

S. Takano, T. Murakami, K. Samizu and K. Ogasawara, *Heterocycles*, **1994**, *39*, 67.

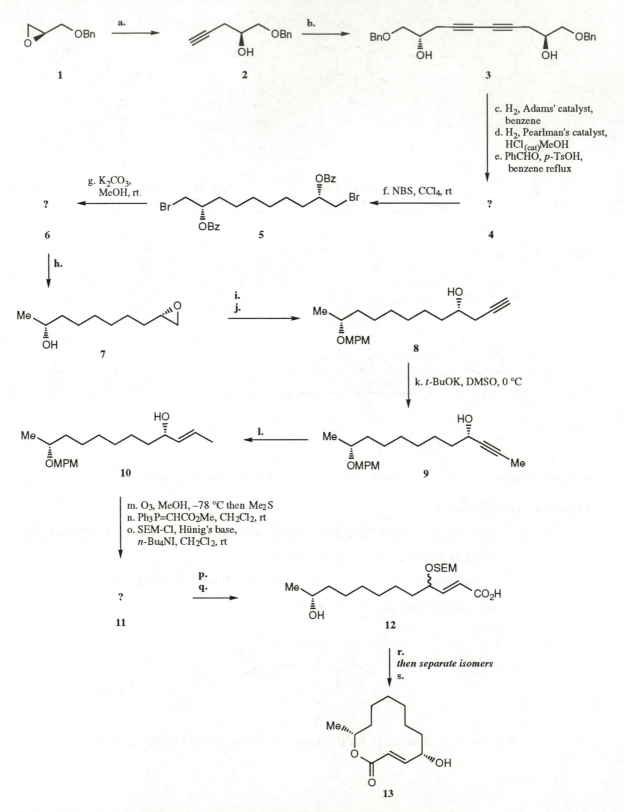

c. H$_2$, Adams' catalyst,
   benzene
d. H$_2$, Pearlman's catalyst,
   HCl$_{(cat)}$MeOH
e. PhCHO, *p*-TsOH,
   benzene reflux

f. NBS, CCl$_4$, rt

g. K$_2$CO$_3$,
   MeOH, rt.

k. *t*-BuOK, DMSO, 0 °C

m. O$_3$, MeOH, –78 °C then Me$_2$S
n. Ph$_3$P=CHCO$_2$Me, CH$_2$Cl$_2$, rt
o. SEM-Cl, Hünig's base,
   n-Bu$_4$NI, CH$_2$Cl$_2$, rt

r.
then separate isomers
s.

Discussion Points

- What is the mechanism of steps **b** and **k**?
- Give a method for the formation of the Z-olefin **14** from alkyne **9**.

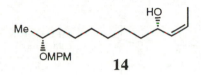

14

Further Reading

- A similar approach to that described above was used by the authors in the synthesis of a milbemycin K fragment, see: S. Takano, Y. Sekiguchi and K. Ogasawara, *Heterocycles*, **1994**, *38*, 59.
- For an excellent review on the two-directional chain synthesis strategy see: C. S. Poss and S. L. Schreiber, *Acc. Chem. Res.*, **1994**, *27*, 9. Another elegant application of this strategy to the synthesis of (–)-parviflorin has recently been published: T. R. Hoye and Z. Ye, *J. Am. Chem. Soc.*, **1996**, *118*, 1801.
- For a review of macrocyclic ring formation methods see: Q. C. Meng and M. Hesse in *Top. Curr. Chem.*, **1992**, *161*, 107.
- A more recent chemoenzymatic synthesis of the related (*R*)-patulolide A has also been published: A. Sharma, S. Sankaranarayanan and S. Chattopadhyay, *J. Org. Chem.*, **1996**, *61*, 1814.

5. Asymmetric Synthesis of 1-Deoxy-8,8a-di-epi-castanospermine

S. F. Martin, H.-J. Chen and V. M. Lynch, *J. Org. Chem.*, **1995**, *60*, 276.

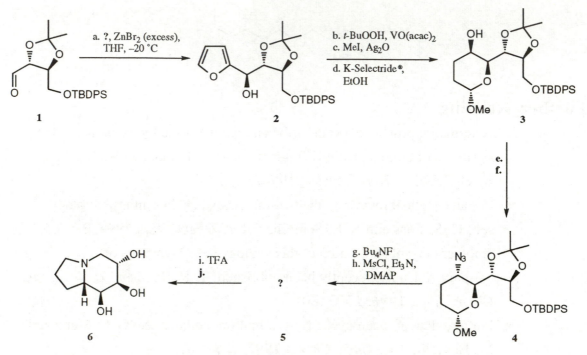

Discussion Points

- Explain the stereoselectivity obtained in step **a**.
- Propose a mechanism for the conversion of **2** into **3**.
- Suggest a structure for the intermediate formed in step **j**.

Further Reading

- For oxidative opening of furans, see: B. M. Adger, C. Barrett, J. Brennan, M. A. McKervey and R. W. Murray, *J. Chem. Soc., Chem. Commun.,* **1991**, *21*, 1553.
- For an analysis of chelation-controlled carbonyl addition reactions, see: M. T. Reetz, *Acc. Chem. Res.*, **1993**, *26*, 462.

6. Synthesis of a Structure Related to Hydantocidin

S. Hanessian, J.-Y. Sancéau and P. Chemla, *Tetrahedron*, **1995**, *51*, 6669.

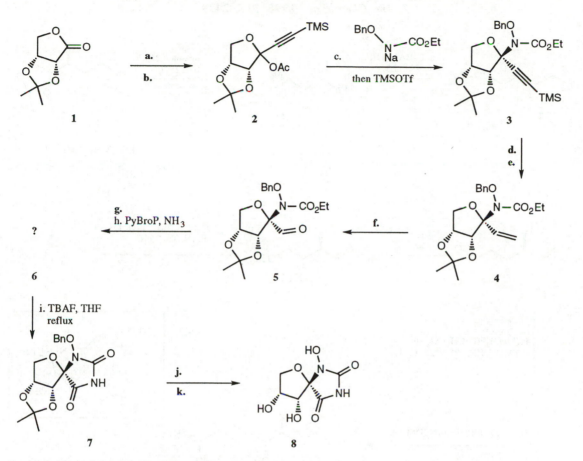

Discussion Points

- Propose a mechanism for step **c** with a rationalisation of the stereochemical outcome.
- What is the role of TBAF in step **i**?

Further Reading

- For recent examples of the use of fluoride ion as a base, see: T. Sato and J. Otera, *J. Org. Chem.*, **1995**, *60*, 2627; see also: T. Sato and J. Otera, *Synlett*, **1995**, 845.

7. Total Synthesis of Cryptophycin C

R. A. Barrow, T. Hemscheidt, J. Liang, S. Paik, R. E. Moore and M. A. Tius,

J. Am. Chem. Soc., **1995**, *117*, 2479.

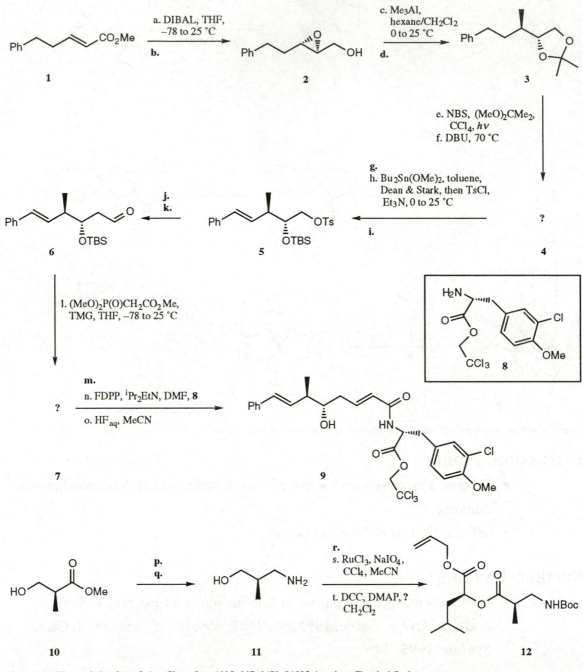

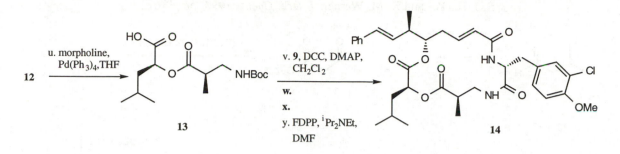

12 → (u. morpholine, Pd(Ph₃)₄, THF) → **13**

13 → (v. **9**, DCC, DMAP, CH₂Cl₂; w.; x.; y. FDPP, ⁱPr₂NEt, DMF) → **14**

Discussion Points

- Suggest a synthesis for compound **1**.
- Explain the regioselectivity in the epoxide opening of step **c**.
- A considerable amount of compound **15** was formed when step **e** was carried out in the absence of 2,2-dimethoxypropane.

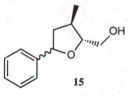

15

  Suggest an explanation for the formation of this compound and the function of 2,2-dimethoxypropane.
- Give mechanisms for steps **h** and **u**.

Further Reading

- For studies on the regioselectivity in epoxide opening, see: M. Chini, P. Crotti, L. A. Flippin, C. Gardelli, E. Giovani, F. Macchia, M. Pineschi, *J. Org. Chem.*, **1993**, *58*, 1221; M. Chini, P. Crotti, C. Gardelli and F. Macchia, *Tetrahedron*, **1994**, *50*, 1261.
- For a review on ring-closure methods in the synthesis of natural products, see: Q. C. Meng and M. Hesse, *Top. Curr. Chem.*, **1992**, *161*, 107.
- For a modification of the allyl group removal procedure, see: A. Merzouk and F. Guibè, *Tetrahedron Lett.*, **1992**, 33, 477.

8. Total Synthesis of (±)-Leuhistin

S. J. Hecker and K. M. Werner, *J. Org. Chem.*, **1993**, *58*, 1762.

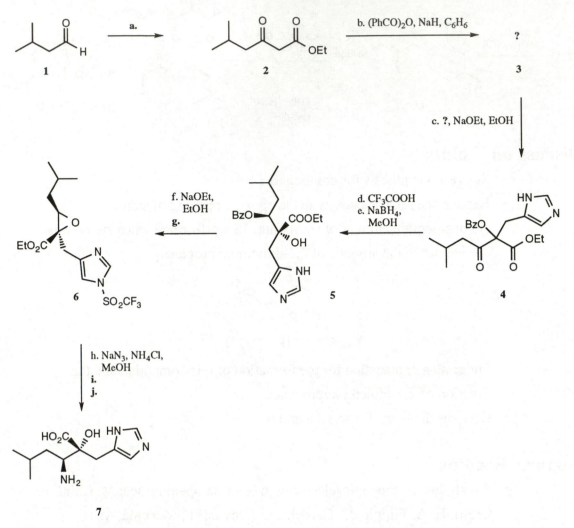

Discussion Points

- Suggest a possible reason for the selectivity observed in step **e**.
- Explain the migration of the benzoate group observed in transforming compound **4** into **5**.

Further Reading

- For the use of activated alumina in the synthesis of 3-oxoesters, see: D. D. Dhavale, P. N. Patil and R. S. Raghao, *J. Chem. Res., Synop.*, **1994**, *4*, 152.
- For reviews on catalytic transfer hydrogenation, see: R. A. Johnstone, A. H. Wilby and I. D. Entwistle, *Chem. Rev.*, **1985**, *85*, 129; G. Brieger and T. J. Nestrick, *Chem. Rev.*, **1974**, *74*, 567.

9. The Asymmetric Synthesis of Bryostatin Fragments

J. De Brabander and M. Vandewalle, *Synthesis*, **1994**, 855.

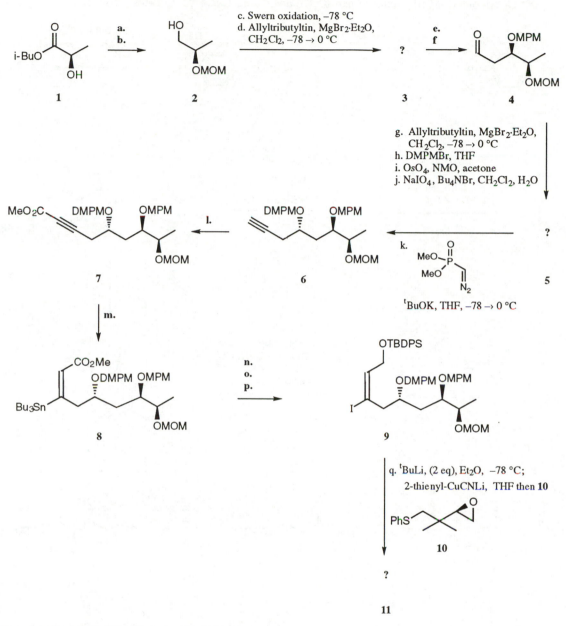

Abstracted with permission from *Synthesis*, **1994**, 855 ©1994 Georg Thieme Verlag

Discussion Points

- What is the purpose of the magnesium bromide etherate in step **d**?
- Propose an explanation for the observed stereochemical outcome of step **g**.
- Explain the role of the Seyferth reagent in the formation of product **6**. What is the mechanism of this step?

13

10. The Total Synthesis of (−)-Balanol

J. W. Lampe, P. F. Hughes, C. K. Biggers, S. H. Smith and H. Hu, *J. Org. Chem.*, **1994**, *59*, 5147.

K. C. Nicolaou, M. E. Bunnage and K. Koide, *J. Am. Chem. Soc.*, **1994**, *116*, 8402

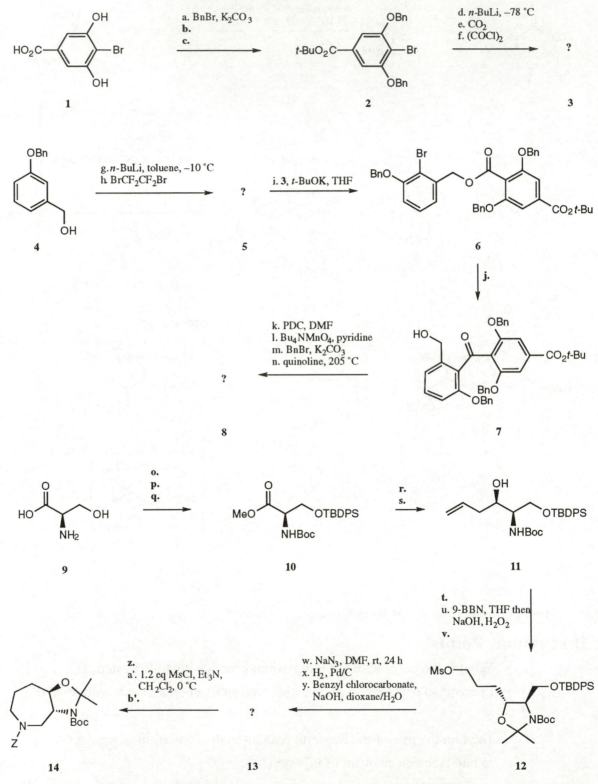

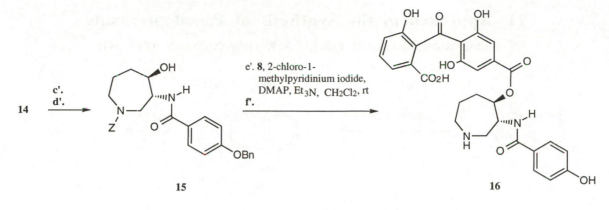

Discussion Points

- Rationalise the selectivity of the transformation of compound **4** into compound **5**.
- What are the usual methods for deprotecting a *tert*-butyl ester?
- Give an alternative method of converting an alcohol into an azide.
- The cyclisation to give the hexahydroazepine **14** was carried out at 'moderate dilution' (0.02 M). What was the reason for this?

Further Reading

- For other total syntheses of balanol and closely related analogues see; T. Naito, M. Torieda, K. Tajiri, I. Ninomiya and T. Kiguchi, *Chem. Pharm. Bull.*, **1996**, *44*, 624; E. Albertini, A. Barco, S. Benetti, C. De Risi, G. Pollini and V. Zanirato, *Synlett*, **1996**, 29; K. C. Nicolaou, K. Koide and M. E. Bunnage, *Chem. Eur. J.*, **1995**, *1*, 454; K. Koide, M. E. Bunnage, L. G. Paloma, J. R. Kanter, S. S. Taylor, L. L. Brunton and K. C. Nicolaou, *Chem. Biol.*, **1995**, *2*, 601-8; C. P. Adams, S. M. Fairway, C. J. Hardy, D. E. Hibbs, M. B. Hursthouse, A. D. Morley, B. W. Sharp, N. Vicker and I. Warner, *J. Chem. Soc. Perkin Trans. 1*, **1995**, 2355; D. Tanner, A. Almario and T. Hoegberg, *Tetrahedron*, **1995**, *51*, 6061.
- For a series of seven articles concerning the biological activity of balanol derivatives, in particular their protein kinase C inhibition, see: J. S. Mendoza, L. Yen-Shi, G. E. Jagdman Jr, W. Lampe *et al.*, *Bioorg. Med. Chem. Lett.*, **1995**, *5*, pages 1839, 2015, 2133, 2147, 2151, 2155, 2211.
- The asymmetric synthesis of the 3-hydroxylysine used by Lampe *et al.* using Sharpless' AD methodology can be found in P. F. Hughes, S. H. Smith and J. T. Olson, *J. Org. Chem.*, **1994**, *59*, 5799.
- For the application of a similar rearrangement to that shown in step **j** see: G. E. Keck, S. F. McHardy and J. A. Murray, *J. Am. Chem. Soc.*, **1995**, *117*, 7289.

11. Approach to the Synthesis of Pseudopterosins

S. W. McCombie, C. Oritz, B. Cox and A. K. Ganguly, *Synlett*, **1993**, 541.

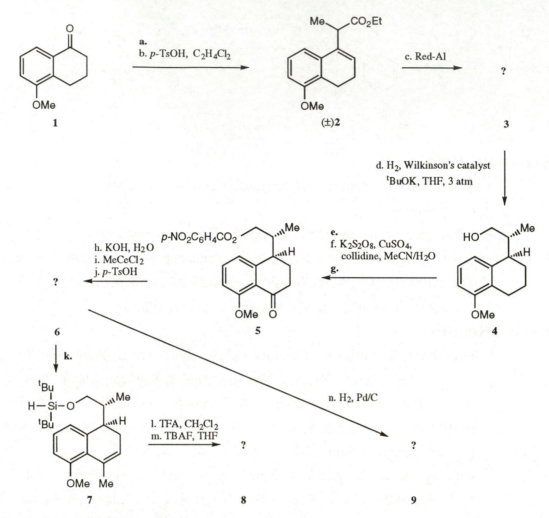

Discussion Points

* What is the purpose of the potassium *t*-butoxide in step **d**? What effect does the choice of Wilkinson's catalyst have over the stereochemical outcome of the reaction?
* What is the mechanism of step **l**?

Further Reading

* For a review on hydrosilylation see: I. Ojima in *The Chemistry of Organic Silicon Compounds* (Ed. S. Patai and Z. Rappoport), Vol. 2, J. Wiley and Sons, Chichester, 1989, pp. 1479-1526.
* Further studies on the synthesis of pseudopterosin A have been published recently: L. Eklund, I. Sarvary and T. Frejd, *J. Chem. Soc., Perkin. Trans. 1*, **1996**, 303.

12. Total Synthesis of Neodolabellenol

D. R. Williams and P. J. Coleman, *Tetrahedron Lett.*, **1995**, *36*, 35.

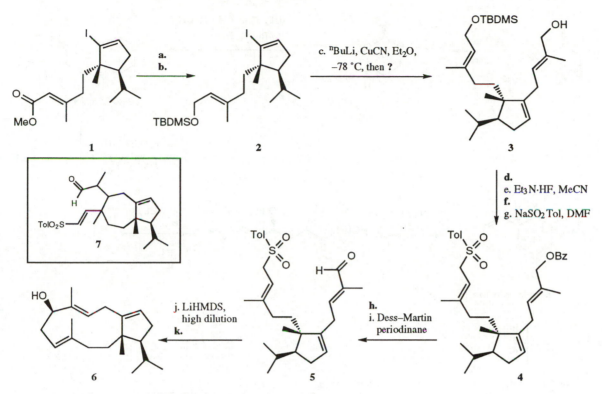

Discussion Points

- Propose a synthesis of **1** starting from 2-methyl-2-cyclopentenone.
- The direct use of the vinyl lithium species generated from **2** preferentially afforded the Z-allylic alcohol in the coupling process **c**. Explain the change in stereoselectivity.
- Give an explanation for the preferential formation of the β-epimer in the Julia coupling **j**.
- A small amount of compound **7** was isolated from the Julia coupling. Give an explanation for its formation.

Further Reading

- For another application of the Julia coupling to the synthesis of macrocycles, see: K. Takeda, A. Nakajima and E. Yoshii, *Synlett*, **1995**, *3*, 249.
- For various aspects of organocuprate-mediated epoxide opening, see: H. Bruce, R. S. Wilhelm, J. A. Kozlowski and D. Parker, *J. Org. Chem.*, **1984**, *49*, 3928.

13. Total Synthesis of (−)-Swainsonine

M. Naruse, S. Aoyagi and C. Kibayashi, *J. Org. Chem.*, **1994**, *59*, 1358.

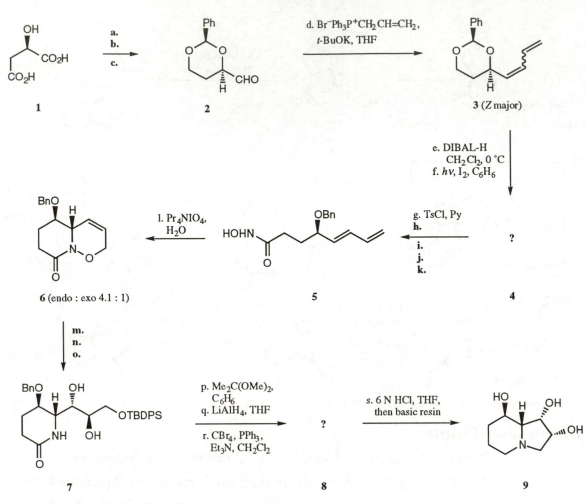

Discussion Points

- Explain the selectivity of the hetero Diels–Alder reaction of step **l** and suggest a reason for the lower selectivity (endo:exo 1.3:1) observed when the reaction was run in chloroform.

- Explain the selectivity observed in the osmylation step **o**.

Further Reading

- For reviews on the use of the hetero Diels–Alder reaction in natural product synthesis, see: J. Streith and A. Defoin, *Synthesis*, **1994**, *11*, 1107; S. F. Martin, *J. Heterocycl. Chem.*, **1994**, *31*, 679; H. Waldmann, *Synthesis*, **1994**, *6*, 535.

14. Total Synthesis of Octalactins A and B

J. C. McWilliams and J. Clardy *J. Am. Chem. Soc.*, **1994**, *116*, 8378.

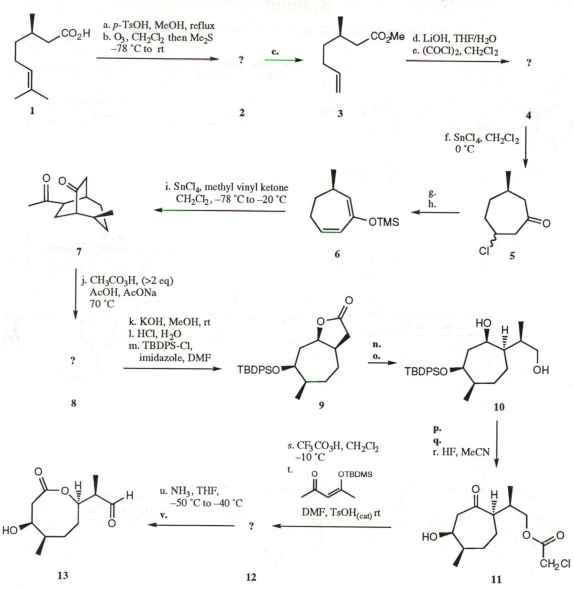

Discussion Points

- Propose a mechanism for the cyclisation step **f**. Why is the seven-membered cyclisation product **5** favoured over a six-membered ring system?
- What is the purpose of the tin tetrachloride in step **i**?
- Explain the regioselectivity obtained in steps **j** and **s**. Rationalise the stereoselectivity of these oxidative ring expansion steps.
- Give a motive for the acidic conditions adopted in step **t**.

Further Reading

- For an extensive review on the Baeyer–Villiger reaction see: G. R. Krow, *Org. React.*, **1993**, *43*, 251.

15. Asymmetric Synthesis of the Milbemycin β₃ Spiroketal Subunit

M. A. Holoboski and E. Koft, *J. Org. Chem.*, **1992**, *57*, 965.

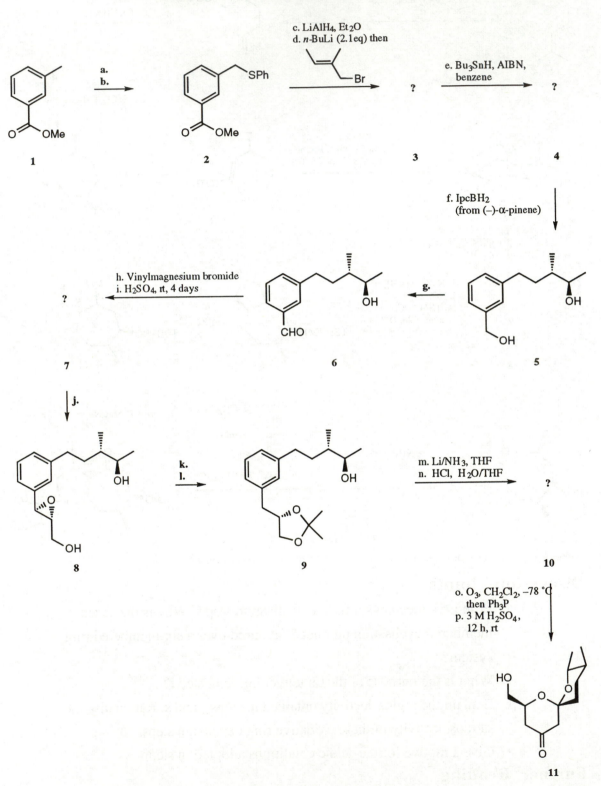

Abstracted with permission from *J. Org. Chem.*, **1992**, *57*, 965. ©1992 American Chemical Society

Discussion Points

- Suggest a method of determining the enantiomeric excess of the product obtained in step **f**.
- What governs the regioselectivity of step **f**?
- Propose a structure for the product of step **h** and the driving force behind its conversion into compound **7**.
- Based on mechanistic considerations and the fact that it displays only two ^{1}H NMR signals between 6.0 ppm and 4.5 ppm, suggest a structure for the product of step **m**.
- What is the mechanism of step **p**? What governs the stereochemistry of the spiroketal centre?

Further Reading

- For a review of approaches to the synthesis of avermectins and milbemycins see: T. A. Blizzard, *Org. Prep. Proced. Int.,* **1994**, *26*, 617.
- For a recent total synthesis of milbemycin G see: S. Bailey, A. Teerawutgulrag and E. J. Thomas, *J. Chem. Soc. Chem. Commun.,* **1995**, 2519 *and* 2521.
- For the synthesis of the spiroacetal fragment of milbemycin β_1 see: S. Naito, M. Kobayashi and A. Saito, *Heterocycles*, **1995**, *41*, 2027.
- For an excellent introduction to hydroboration see S. E. Thomas, *Organic Synthesis — The Roles of Boron and Silicon*, Oxford University Press, Oxford, 1991, pp.1–8. A more detailed description can be found in: A. Pelter, K. Smith and H. C. Brown, *Borane Reagents*, Academic Press 1988, pp.165–230.
- For another application of a Birch reduction–ozonolysis sequence see: G. Zvilchovsky and V. Gury, *J. Chem. Soc. Perkin Trans. 1*, **1995**, 2509.
- For structural details of more recently isolated milbemycins see: G. H. Baker, S. E. Blanchflower, R. J. Dorgan, J. R. Everett, B. R. Manger, C. R. Reading, S. A. Readshaw and P. Shelley. *J. Antibiot.,* **1996**, *49*, 272.

16. Stereoselective Total Synthesis of (+)-Artemisinin

M.A. Avery, W.K.M Chong and C. Jennings-White, *J. Am. Chem. Soc.*, **1992**, *114*, 974.

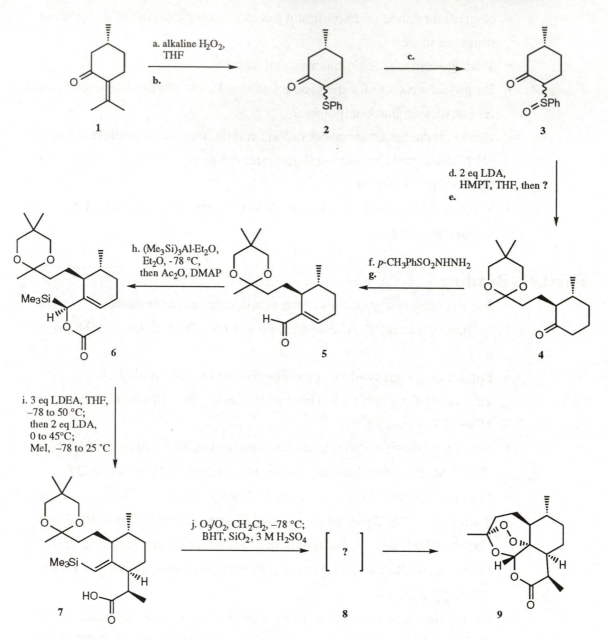

Discussion Points

- Suggest a motive for oxidising compound **2** before the alkylation step **d**.
- Explain the stereoselectivity observed in step **h**?
- What other ester could have directly afforded compound **7** via a Claisen rearrangement?
- Suggest a mechanism for the formation of compound **8**, obtained upon ozonolysis of vinylsilane **7**.

Further Reading

- For an analysis of recent syntheses of artemisinin, see: M. G. Constantino, M. Beltrame and G. V. J. Dasilva, *Synth. Commun.*, **1996**, *26*, 321; W.-S. Zhou and X.-X. Xu, *Acc. Chem. Res.*, **1994**, *27*, 211–216.

- For syntheses of structural analogues of artemisinin, see: G. H. Posner, C. H. Oh, L. Gerena and W. K. Milhous, *J. Med. Chem.*, **1992**, *35*, 2459.

- For studies on the mechanism of action of artemisinin, see: G. H. Posner, J. N. Cumming, P. Ploypradith and C. H. Oh, *J. Am. Chem. Soc.*, **1995**, *117*, 5885.

- For a review on applications of the Shapiro reaction, see: R. M. Adlington and A. G. M. Barrett, *Acc. Chem. Res.*, **1983**, *16*, 55.

- For a review of the Claisen rearrangement, see: S. Pereira and M. Srebnik, *Aldrichimica Acta*, **1993**, *26*, 17.

- For the use of β-keto sulphoxides in synthesis, see: P. A. Grieco and C. S. Pogonowski, *J. Org. Chem.*, **1974**, *39*, 732.

17. The Synthesis of (±)-Prosopinine

G. R. Cook, L. G. Beholz and J. R. Stille, *Tetrahedron Lett.*, **1994**, *35*, 1669.

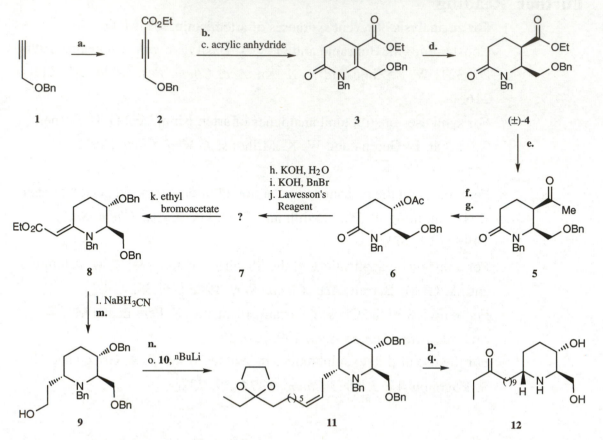

Discussion Points

- Suggest a mechanism for the annulation of alkyne **2** into the δ-lactam **3**.
- What product could be expected from the direct Baeyer–Villiger oxidation of compound **5**?
- Propose a structure for compound **10**.
- Compound **11** was formed in an 85 : 15 mixture with its *trans*-isomer. Explain the selectivity of the reaction.

Further Reading

- For reviews on the stereochemical aspects of the Wittig and related reactions see: E. Vedejs and M. J. Peterson, *Topics in Stereochemistry*, **1994**, *21*, 1; B. E. Maryanoff and A. B. Reitz, *Chem. Rev.*, **1989**, *89*, 863.
- For an extensive review of the thionation reactions of Lawesson's reagent see: M. P. Cava and M. I. Levinson, *Tetrahedron*, **1985**, *41*, 5061.
- For work related to this article see also: G. R. Cook, L. G. Beholz and J. R. Stille, *J. Org. Chem.*, **1994**, *59*, 3575.

18. Synthesis of a Protected Fluorocarbocyclic Nucleoside

I. C. Cotterill, P. B. Cox, A. F. Drake, D. M. Le Grand, E. J. Hutchinson, R. Latouche, R. B. Pettman, R. J. Pryce, S. M. Roberts, G. Ryback, V. Sik, and J. O. Williams , *J. Chem. Soc. Perkin Trans. 1*, **1991**, 3071.

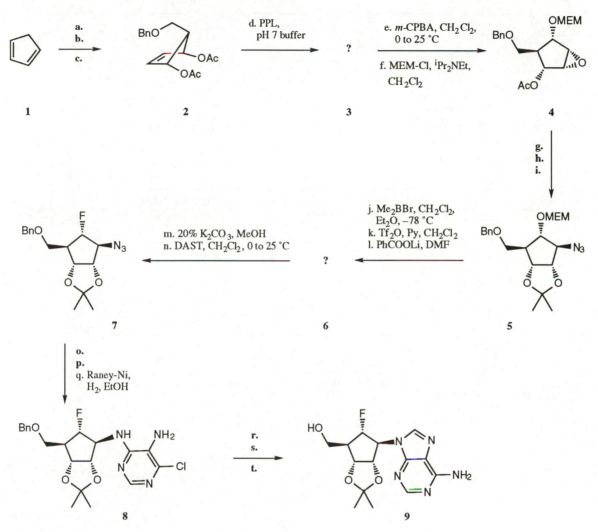

Abstracted with permission from *J. Chem. Soc. Perkin Trans. 1*, **1991**, 3071 ©1991 The Royal Society of Chemistry

Discussion Points

- Give reasons for the stereoselectivity of epoxidation step **e**.
- What is the mechanism of the fluorination step **n**?

Further Reading

- For a review on selectivity in enzymatic reactions, see: J. B. Jones, *Aldrichimica Acta*, **1993**, *26*, 105. For other studies on active site model for lipases, see: K. Naemurz, R. Fukuda, M. Murata, M. Konishi, K. Hirose and Y. Tobe, *Tetrahedron: Asymm.*, **1995**, *6*, 2385.
- For a review on the fluorination of organic compounds, see: O. A. Mascaretti, *Aldrichimica Acta*, **1993**, *26*, 47.

19. Synthesis of Tropane Alkaloid Calystegine A3

C. R. Johnson and S. J. Bis, *J. Org. Chem.*, **1995**, *60*, 615.

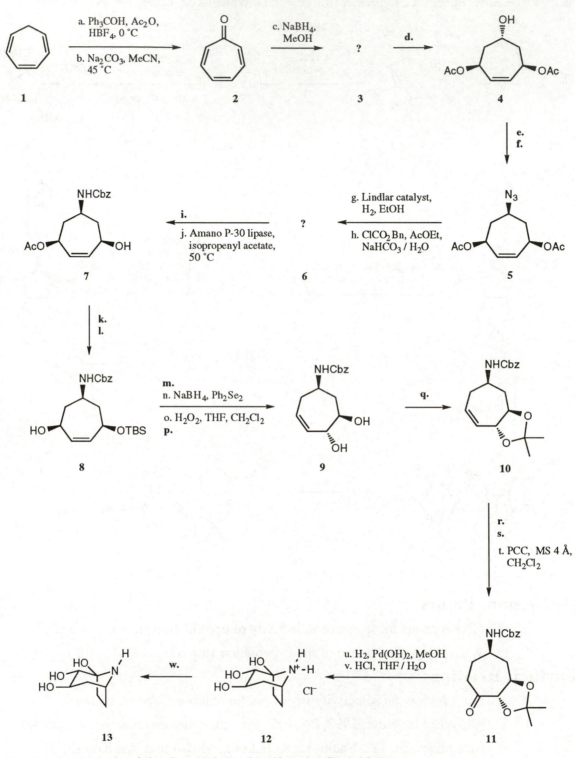

Abstracted with permission from *J. Org. Chem.*, **1995**, *60*, 615 ©1995 American Chemical Society

Discussion Points

- What is the mechanism for the formation of compound **2**?
- Explain the diastereoselectivity of the palladium-catalysed diacetoxylation of compound **3**.
- Suggest an enzymatic route to *ent*-**7** from compound **6**.
- The enzymatic asymmetrisation carried out on the azido derivative **5** or its diol derivative did not meet with success. Give a possible reason for the low enantioselectivity in this case.
- What is the mechanism for the formation of diol **9**?
- A considerable amount of compound **14** is formed in the hydroboration oxidation steps **r, s**. Propose a possible explanation why there should be no strong directing effect of the α–oxygen in this case.

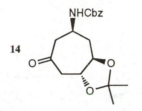

Further Reading

- For references on the stereo- and regioselectivity in the palladium-catalysed diacetoxylation of dienes, see: J. E. Bäckvall, S. E. Bystroem and R. E. Nordberg, *J. Org. Chem.*, **1984**, *49*, 4619; J. E. Bäckvall and R. E. Nordberg, *J. Am. Chem. Soc.,* **1981**, *103*, 4959.
- For the combined use of enzymatic transformation and palladium chemistry in asymmetric synthesis, see: J. E. Bäckvall, R. Gatti and H. E. Schink, *Synthesis*, **1993**, *3*, 343; J. V. Allen and J. M. Williams, *Tetrahedron Lett.*, **1996**, *37*, 1859.
- For the use of enzymes in organic solvents, see: A. M. Klibanov, *Acc. Chem. Res.*, **1990**, *23*, 114; C.-S. Chen and C. J. Sih, *Angew. Chem.*, **1989**, *101*, 711.
- For the enzymatic transformation of tropone derivatives into sugars and related products, see: C. R. Johnson, A. Golebiowski, D. H. Steensma and M. A. Scialdone, *J. Org. Chem.*, **1993**, *58*, 7185.

20. The Synthesis of (±)-Oxerine

Y. Aoyagi, T. Inariyama, Y. Arai, S. Tsuchida, Y. Matuda, H. Kobayashi, A. Ohta, T. Kurihara and S. Fujihira, *Tetrahedron*, **1994**, *50*, 13575.

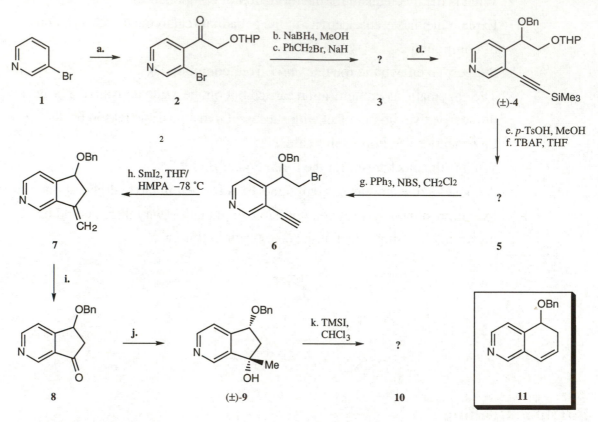

Discussion Points

- What is the predominant reason for the selectivity of the alkylation step **a**?
- Propose a mechanism for step **d**.
- Propose a mechanism for the samarium iodide promoted cyclisation of acetylene **6**. What is the reactive intermediate responsible for this step? Give reasons why the formation of cyclohexene **11** should be disfavoured.
- Why was it not possible to obtain alcohol **9** directly by hydroboration of olefin **7**?
- What effect is responsible for the high diastereoselectivity observed in the methylation of ketone **8**?

Further Reading

- For a review on Baldwin's rules, see: C. D. Johnson, *Acc. Chem. Res.*, **1993**, *26*, 476.
- For a recent review on samarium iodide, see: G. A. Molander and C. R. Harris, *Chem. Rev.*, **1996**, *96*, 307.

21. Synthesis of an Enantiopure
C-4 Functionalised 2-Iodocyclohexanone Acetal

Z. Su and L. A. Paquette, *J. Org. Chem.*, **1995**, *60*, 764.

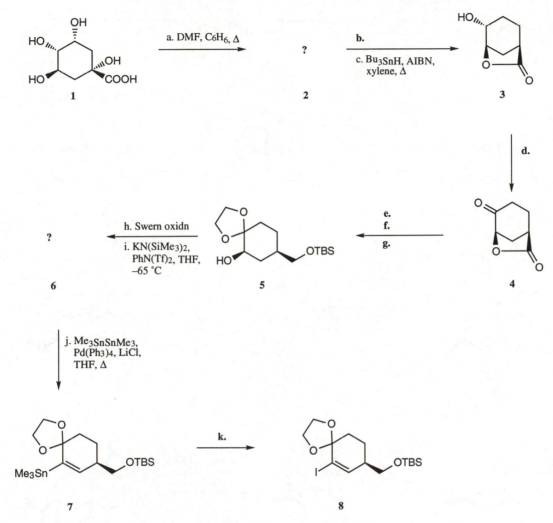

Discussion Points

- Explain the selectivity in the formation of the lactone ring in step **a**.
- What is the purpose of lithium chloride in step **j**?

Further Reading

- For a modification of the Barton deoxygenation reaction using polystyrene-supported organotin hydride, see: W. P. Neumann and M. Peterseim, *Synlett*, **1992**, *10*, 801. For a review on the Barton–McCombie
  methodology, see: C. Chatgilialoglu and C. Ferreri, *Res. Chem. Intermed.*, **1993**, *19*, 755.
- For a review on the use of vinyl triflates in synthesis, see: K. Ritter, *Synthesis*, **1993**, *8*, 735.

22. Total Synthesis of (–)-Solavetivone

J. R. Hwu and J. M. Wetzel, *J. Org. Chem.*, **1992**, *57*, 922.

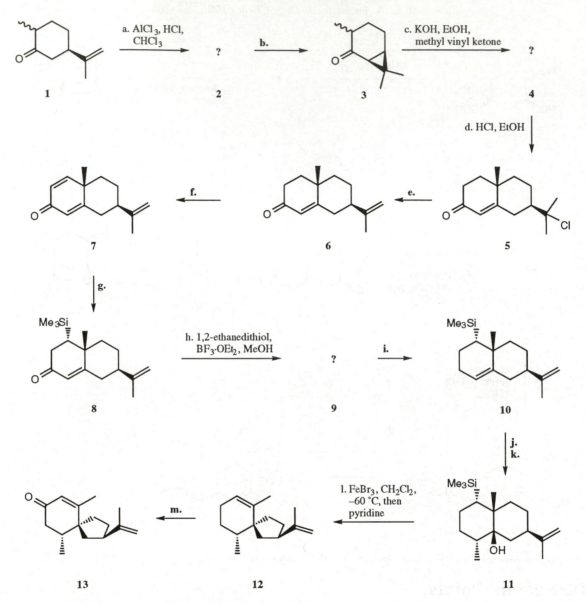

Discussion Points

- Suggest a mechanism for step **d**.
- The conversion of compound **6** into the trienone **7** was carried out in a single oxidative step. Propose alternative methods for the introduction of α,β unsaturation in a ketone.
- What factors govern the regioselectivity and stereoselectivity of the conjugate additon of the trimethylsilyl nucleophile in step **g**?
- A homonuclear NMR decoupling experiment was carried out on compound **11**, simultaneously irradiating the two proton signals at δ 4.69 and 4.70 ppm. This resulted in the simplification of the single proton

multiplet between δ 2.35 – 2.87 ppm. In a subsequent nOe experiment an 18% enhancement of the aforementioned multiplet was observed on irradiation of a methyl group doublet at δ 1.26 ppm. Rationalise the above observations in terms of the structure and stereochemistry given for **11**.

- Suggest a mechanism for the Lewis acid catalysed rearrangement in step **l**.

Further Reading

- For other recent strategies for spirocyclisation see: D. L. J. Clive, X. Kong and C. C. Paul, *Tetrahedron*, **1996**, *52*, 6085; A. Srikrishna, P. P. Praveen and R. Viswajanani, *Tetrahedron Lett.*, **1996**, *37*, 1683; Y. I. M. Nilsson, A. Aranyos, P. G. Andersson, J.-E. Bäckvall, J.-L. Parrain, C. Ploteau and J-P Quintard, *J. Org. Chem.*, **1996**, *61*, 1825; A. Srikrishna, T. Reddy, K. P. Jagadeeswar and V. D. Praveen, *Synlett*, **1996**, 67; R. Grigg, B. Putnicovic and C. J. Urch, *Tetrahedron Lett.*, **1996**, *37*, 695.

23. Total Synthesis of (+)-Himbacine

D. J. Hart, W.-L. Wu and A. P. Kozikowski, *J. Am. Chem. Soc.*, **1995**, *117*, 9369.

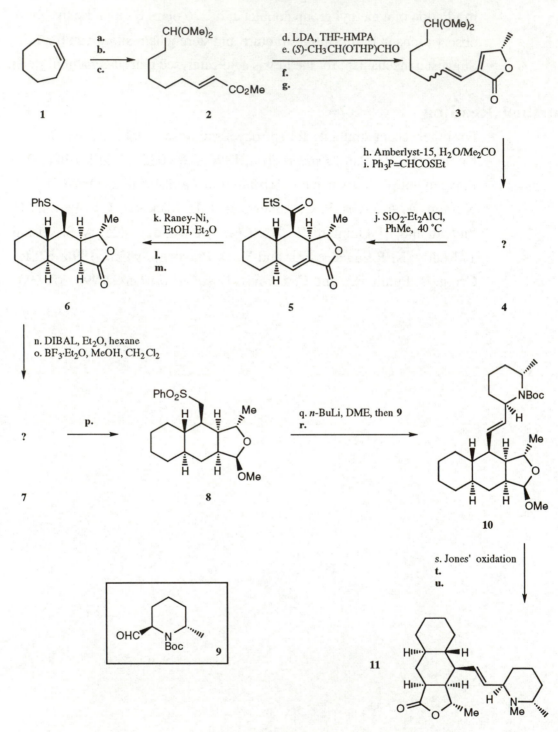

Discussion Points

- The reaction sequence starting from the dienolate derived from **2** led to a 8:1 mixture of the *E:Z* isomers corresponding to **3**. Suggest a method to improve the stereoisomeric ratio.

- A significantly lower selectivity in the Diels–Alder reaction was obtained with the use of the alkyl ester corresponding to **4** or when no catalyst was used. Give reasons for this finding.

- Suggest a starting material for the synthesis of aldehyde **9**.

Further Reading

- For the control in the regioselective alkylation of dienolates, see: Y Yamamoto, S. Hatsuya and J. Yamada, *J. Org. Chem.*, **1990**, *55*, 3118.

- For the use of samarium diiodide in the Julia-Lythgoe olefination, see: G. E. Keck, K. A. Savin and M. A. Weglarz, *J. Org. Chem.*, **1995**, *60*, 3194.

- For some reviews on the intramolecular Diels–Alder reaction, see: W. R. Roush, *Adv. Cycloadd.*, **1990**, *2*, 91; M. E. Jung, *Synlett*, **1990**, *4*, 186; D. Craig, *Chem. Soc. Rev.*, **1987**, *16*, 187; A. G. Fallis, *Can. J. Chem.*, **1984**, *62*, 183.

- For a review on the enhancement of rate and selectivity in Diels–Alder reactions, see: U. Pindur, G. Lutz and C. Ott, *Chem. Rev.*, **1993**, *93*, 741.

24. Synthesis of 5-Hydroxytiagabine

K. E. Andersen, M. Begtrup, M. S. Chorghade, E. C. Lee, J. Lau, B. F. Lundt, H. Petersen, P. O. Sørensen and H. Thøgersen, *Tetrahedron*, **1994**, *50*, 8699.

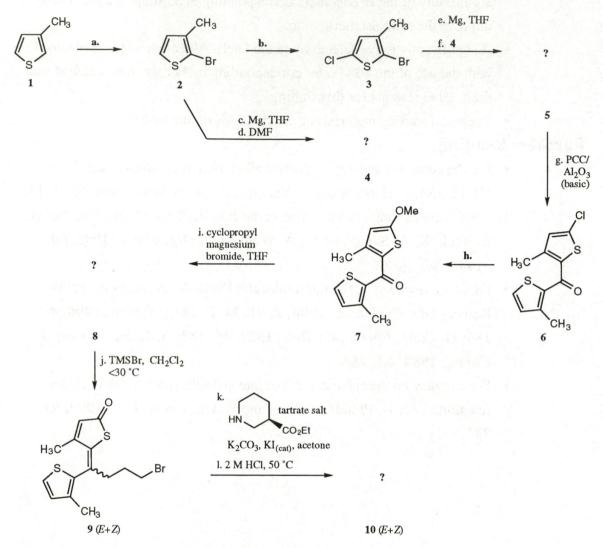

Discussion Points

- What governs the selectivity of the bromination of 3-methylthiophene **1**?
- Propose a mechanism for the rearrangements occuring in step **j**.
- What is the purpose of the catalytic amount of potassium iodide used in step **k**?

Further Reading

- For a recent account of the biological profile of tiagabine see: T. Halonen, J. Nissinen, J. A. Jansen and A. Pitkaenen, *Eur. J. Pharmacol.*, **1996**, *299*, 69.

25. An Approach to Zoanthamine Alkaloids

D. Tanner, P. G. Andersson, L. Tedenborg and P. Somfai, *Tetrahedron*, **1994**, *50*, 9135.

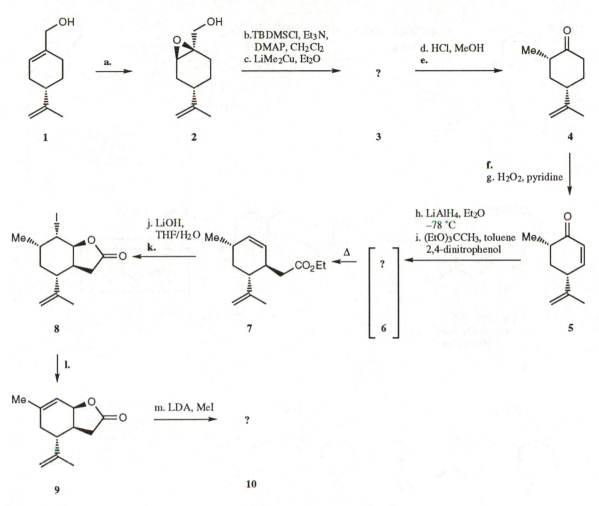

Discussion Points

- What was the stereochemical outcome of the reduction step **h**?
- Propose a mechanism to explain the formation and stereochemistry of compound **7**. What is the structure of the intermediate **6** formed prior to the rearrangement?

Further Reading

- For a recent review of asymmetric and enantioselective epoxidations see: E. Hoeft, *Topp. Curr. Chem.,* **1993**, 63.
- For the application of the Johnson orthoester Claisen rearrangement to a similar system see: A. Srikrishna and R. Viswajanani, *Tetrahedron Lett.,* **1996**, *37*, 2863.

26. Total Synthesis of 1233A

P. M. Wovkulich, K. Shankaran, J. Kigiel and M. R. Uskokovic, *J. Org. Chem.*, **1993**, *58*, 832.

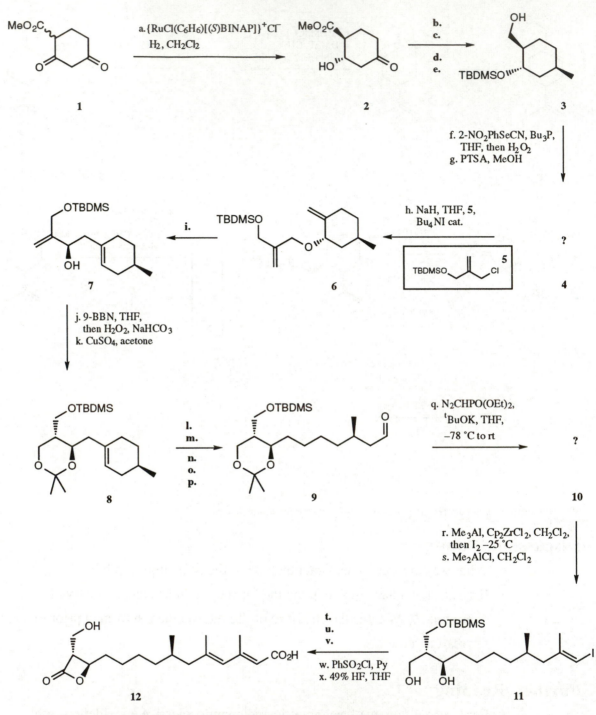

Discussion Points

- What is the mechanism of the Noyori catalyst-mediated asymmetric hydrogenation step **a**?
- Explain the diastereoselectivity observed in the [2,3]-rearrangement of step **i**.
- During the hydroboration step, a substantial amount of compound **13** was formed. Suggest a way to convert it into the desired epimer **8** by exploiting the latent symmetry of the system.

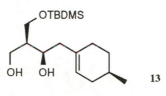

Further Reading

- For the influence of trace amount of acid in the ruthenium (II)-BINAP catalysed asymmetric hydrogenation, see: S. A. King, A. S. Thompson, A. O. King and R. T. Verhoeven, *J. Org. Chem.*, **1992**, *57*, 6689.
- For a review on enantioselective catalysis with transition metal compounds, see: H. Brunner, *Quim. Nova*, **1995**, *18*, 603.
- For reviews on the [2,3]-Wittig sigmatropic rearrangement, see: K. Mikami and T. Nakai, *Synthesis*, **1991**, *8*, 594; K. Mikami and T. Nakai, *Chem. Rev*, **1986**, *86*, 885.
- For a review on the carbometalation of alkynes, see: E. Negishi, *Pure Appl. Chem.*, **1981**, *53*, 2333.

27. Synthesis of a Key Intermediate of 1ß-Methylcarbapenem Antibiotics

S.-H. Kang and H.-S. Lee, *Tetrahedron Lett.*, **1995**, *36*, 6713.

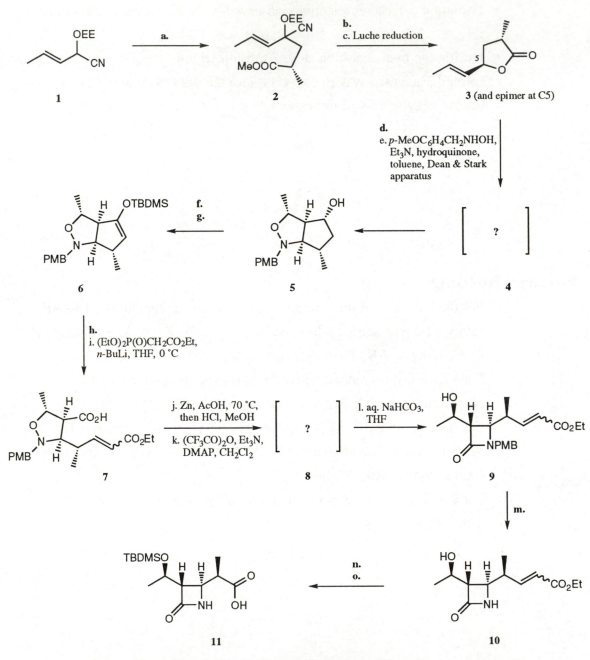

Abstracted with permission from *Tetrahedron Lett.*, **1995**, *36*, 6713 ©1995 Elsevier Science Ltd

Discussion Points

- When *epi-3* was submitted to the sequence outlined in steps **d** and **e**, a mixture of diastereoisomers was formed. Propose their structure and explain the reduced stereoselectivity observed.

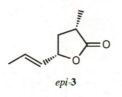

epi-3

- How could *epi-3* be converted into compound **3**?
- What is the purpose of hydroquinone in step **e**?
- Give the structure of the reactive intermediate of step **k**.

Further Reading

- For a review of the use of protected cyanohydrins as acyl anion equivalents, see: J. D. Albright, *Tetrahedron*, **1983**, *39*, 3207.
- For reviews on nitrone cycloadditions to olefins, see: P. N. Confalone and E. M. Huie, *Org. React.*, **1988**, *36*, 1; R. Annunziata, M. Cinquini, F.Cozzi and L. Raimondi, *Gazz. Chim. Ital.*, **1989**, *119*, 253.
- For reviews of the Mitsunobu reaction, see: D. L. Hughes, *Org. React.*, **1992**, *42*, 335; D. L. Hughes, *Organic Preparations and Procedures Int.*, **1996**, *28*, 127.

28. Total Synthesis of the Enantiomer of Hennoxazole A

P. Wipf and S. Lim, *J. Am. Chem. Soc.*, **1995**, *117*, 558.

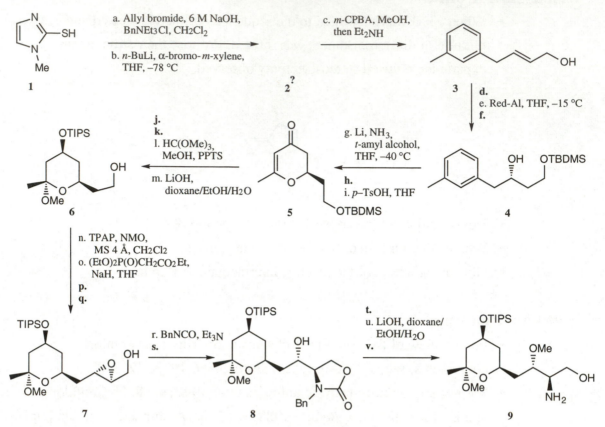

Discussion Points

- What is the mechanism of the Evans–Mislow rearrangement in step **c**?

- Suggest a mechanism for the formation of compound **5**.

- Explain the stereoselectivity observed in the Luche reduction step **j**.

Further Reading

- For a mechanistic explanation of the selectivity of the Sharpless epoxidation, see: E. J. Corey, *J. Org. Chem.*, **1990**, *55*, 1693; P. G. Potvin and S. Bianchet, *J. Org. Chem.*, **1992**, *57*, 6629.

- For reviews on the Birch reduction, see: P. W. Rabideau, *Tetrahedron*, **1989**, *45*, 1579; J. M. Hook and L. N. Mander, *Nat. Prod. Rep.*, **1986**, *3*, 35.

- For a review on the use of tetrapropylammonium perruthenate in organic synthesis, see: S. V. Ley, J. Norman and S. P. Marsden, *Synthesis*, **1994**, *7*, 639.

- For an analysis of the mechanism of the Evans–Mislow rearrangement, see: D. K. Jones-Hertzog and W. L. Jorgensen, *J. Org. Chem.*, **1995**, *60*, 6682; D. K. Jones-Hertzog, W. L. Jorgensen, *J. Am. Chem. Soc.*, **1995**, *117*, 9077.

29. Total Synthesis of (+)-Duocarmycin A

D. L. Boger, J. A. McKie, T. Nishi and T. Ogiku, *J. Am. Chem. Soc.*, **1996**, *118*, 2301.

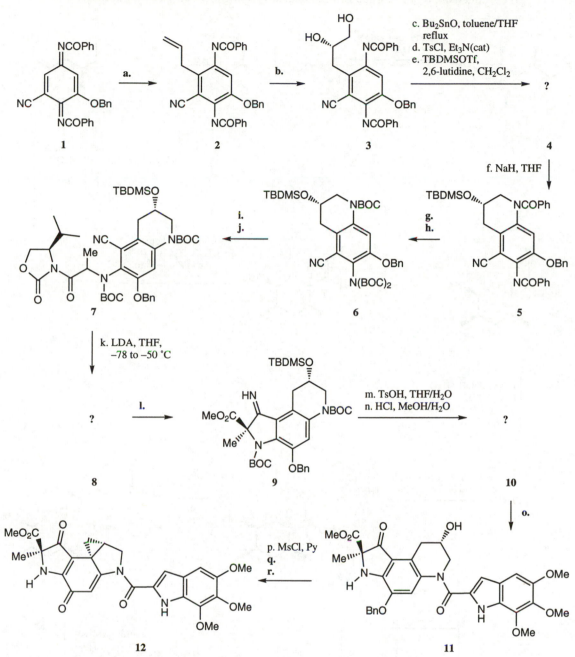

Discussion Points

- Step **k** was carried out by adding the LDA to the substrate. It was found that if the substrate was added to the base, an almost complete inversion of the newly formed quaternary stereocentre α to the imine was obtained. Suggest an explanation for this observation.

Further Reading

- For a review on the use of allylstannanes see: E. J. Thomas, *Chemtracts*, **1994**, *7*, 207.

30. Studies Towards the Total Synthesis of Rapamycin

J. C. Anderson, S. V. Ley and S. P. Marsden, *Tetrahedron Lett.*, **1994**, *35*, 2087.

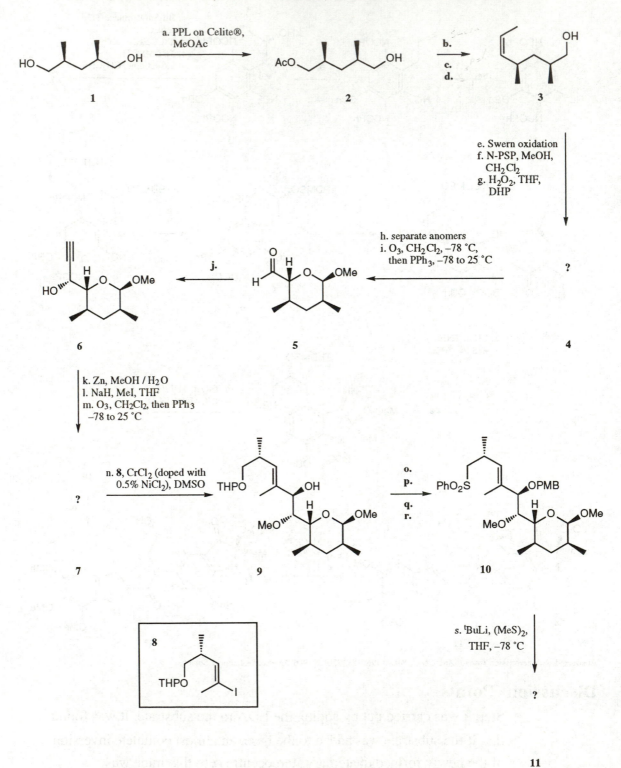

Discussion Points

- What is the maximum theoretical yield for step **a**?
- Give a mechanism for the cyclisation step **f**?
- What is the use of PPh$_3$ in step **i**?
- Explain the selectivity observed in step **j**.
- The anomer of **5** was found to react with lower selectivity. Suggest a reason for this finding.
- Propose a synthesis of vinyl iodide **8**.
- How could the minor, undesired diastereomer obtained in the Nozaki–Kishi condensation of step **n** be converted into **9**?

Further Reading

- For the use of enzymes in organic solvents, see: A. L. Gutman and M. Shapira, *Adv. Biochem. Eng./Biotechnol.*, **1995**, *52*, 87; L. Kvittingen, *Tetrahedron*, **1994**, *50*, 8253; A. P. G. Kieboom, *Biocatalysis*, **1990**, 357; C. H. Wong, *Science*, **1989**, *244*, 1145; C. S. Chen and C. J. Sih, *Angew. Chem.*, **1989**, *101*, 711; A. M. Klibanov, *Trends Biochem. Sci.*, **1989**, *14*, 141.
- For a review on the preparation and reactivity of alkenyl-zinc, -copper, and -chromium organometallics, see: P. Knochel and C. J. Rao, *Tetrahedron*, **1993**, *49*, 29.
- For a review on the Nozaki–Kishi reaction, see: P. Cintas, *Synthesis*, **1992**, 248.
- For the other papers in the series, see: C. Kouklovsky, S. V. Ley and S. P. Marsden, *Tetrahedron Lett.*, **1994**, *13*, 2091; S. V. Ley, J. Norman and C. Pinel, *Tetrahedron Lett.*, **1994**, *35*, 2095.

31. Studies Towards the Total Synthesis of Rapamycin

C. Kouklovsky, S. V. Ley and S. P. Marsden, *Tetrahedron Lett.*, **1994**, *35*, 2091.

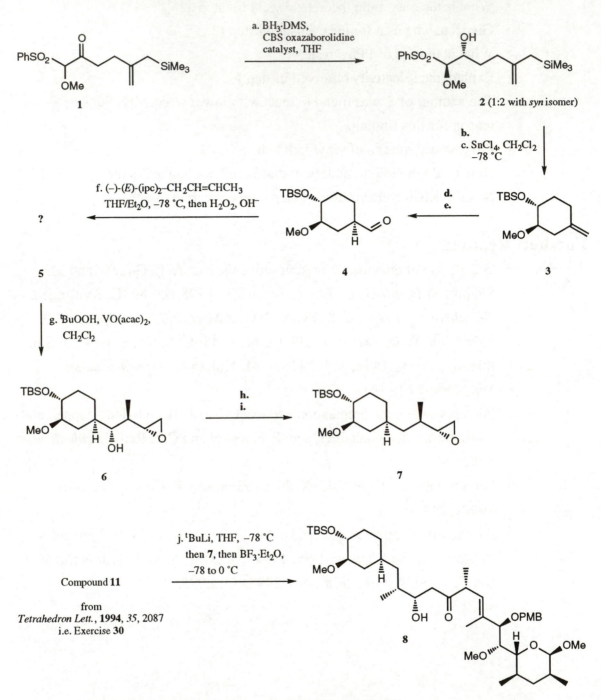

Discussion Points

- How would it be possible to recycle the undesired *syn* isomer obtained in step **a**?
- What is the mechanism of step **c**?
- Explain the diastereoselectivity observed in the (*E*)-crotylborane addition to aldehyde **4**.
- Rationalise the diastereoselectivity observed in the epoxidation of homoallylic alcohol **5**. How could the stereoselectivity of this reaction be reversed?

Further Reading

- For an analysis of the mechanism in the oxazaborolidine catalysed reduction of ketones, see: B. Ganem, *Chemtracts: Org. Chem.,* **1988**, *1*, 40.
- For a review on enantioselective reduction of ketones, see: V. K. Singh, *Synthesis*, **1992**, *7*, 607.
- For a review on the diastereoselectivity in nucleophilic additions to unsymmetrically substituted carbonyls, see: B. W. Gung, *Tetrahedron*, **1996**, *52*, 5263.
- For reviews on the synthesis of chiral epoxides, see: P. Besse and H. Veschambre, *Tetrahedron*, **1994**, *50*, 8885; W. Adam and M. J. Richter, *Acc. Chem. Res.*, **1994**, *27*, 57; K. A. Jørgensen, *Chem. Rev.*, **1989**, *89*, 431.
- For a review on substrate-directable reactions, see: A. H. Hoveyda, D. A. Evans and G. C. Fu, *Chem. Rev.*, **1993**, *93*, 1307.
- For a review on epoxide chemistry, see: J. Gorzynski Smith, *Synthesis*, **1984**, 629.
- For a review on intramolecular reactions of allylic and propargylic silanes, see: D. Schinzer, *Synthesis*, **1988**, *4*, 263.
- For a review on the use of chiral boranes in synthesis, see: D. S. Matteson, *Chem. Rev.*, **1986**, 973.

32. Studies Towards the Total Synthesis of Rapamycin

S. V. Ley, J. Norman and C. Pinel, *Tetrahedron Lett.*, **1994**, *35*, 2095.

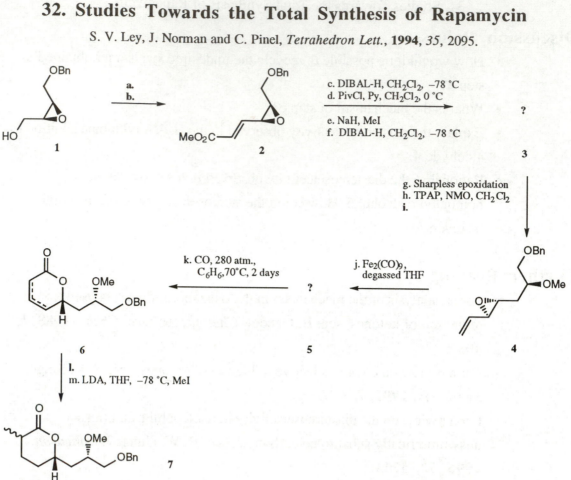

Abstracted with permission from *Tetrahedron Lett.*, **1994**, *35*, 2095 ©1994 Elsevier Science Ltd

Discussion Points

• Which enantiomer of diethyl tartrate is necessary in the Sharpless epoxidation step **g** to obtain the required stereoisomer?

Further Reading

• For the synthesis of lactones via tricarbonyliron–lactone complexes, see: S. V. Ley, L. R. Cox and G. Meek, *Chem. Rev.*, **1996**, *96*, 423; G. D. Gary, S. V. Ley, C. R. Self and R. Sivaramakrishnan, *J. Chem. Soc., Perkin Trans. 1*, **1981**, *1*, 270; R. Aumann, H. Ring, C. Krueger and R. Goddard, *Chem. Ber.*, **1979**, *112*, 3644.

• For reviews on the use of iron carbonyl complexes, see: G. D. Annis, E. M. Hebblethwaite, S. T. Hodgson, A. M. Horton, D. M. Hollinshead, S. V. Ley and R. Sivaramakrishnan, *Spec. Publ. R. Soc. Chem.*, **1984**, *50*, 148; J. Rodriguez, P. Brun and B. Waegell, *Bull. Chem. Soc. Fr.*, **1989**, 799.

33. Synthesis of Isochromanquinones

M. P. Winters, M. Stranberg and H. W. Moore, *J. Org. Chem.*, **1994**, *59*, 7572.

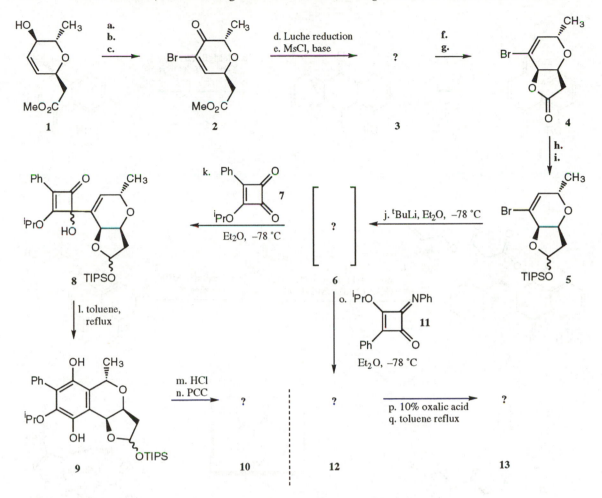

Abstracted with permission from *J. Org. Chem.*, **1994**, *59*, 7572 ©1994 American Chemical Society

Discussion Points

- What is the the mechanism of the bromination steps involved in the synthesis of compound **2**?
- Suggest a structure for the unstable intermediate **6**.
- Suggest a reaction mechanism for the thermal rearrangement of cyclobutenone **8** into the isochromanhydroquinone **9**.

Further Reading

- For a recent study of squarate ester reactivity see: L. A. Paquette and T. Morwick, *J. Am. Chem. Soc.*, **1995**, *117*, 1451.
- For a review, in German, of the reactions of squaric acid and its derivatives see: A. H. Schmidt, *Synthesis*, **1980**, 961.
- For a recent application of this rearrangement to 4-allenylcyclobutanones see: M. Taing and H. W. Moore, *J. Org. Chem.*, **1996**, *61*, 329.

34. The Synthesis of (±)-12a-Deoxytetracycline

G. Stork, J. J. La Clair, P. Spargo, R. P. Nargund and N. Totah, *J. Am. Chem. Soc.*, **1996**, *118*, 5304.

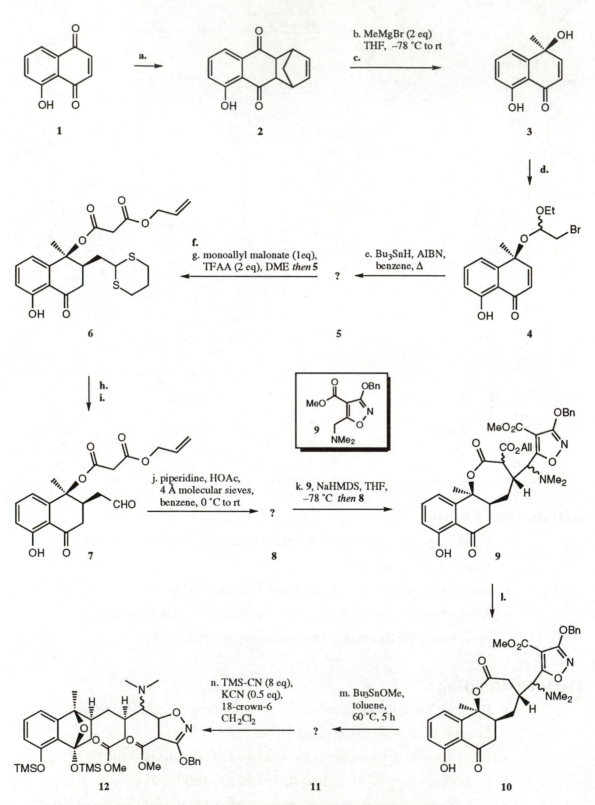

34. The Synthesis of (±)-12a-Deoxytetracycline

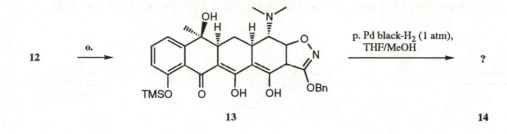

12 —o.→ **13** —p. Pd black-H₂ (1 atm), THF/MeOH→ ? **14**

Discussion Points

- What is the purpose of the Diels–Alder cycloaddition step **a**?
- Give an explanation for the selectivity in the addition of the Grignard reagent in step **b**. How is it dependent on the number of equivalents of methyl magnesium bromide employed?
- What is the mechanism of step **e**?
- What governs the stereoselectivity of the addition of the anion of isoxazole **9** to compound **8**?

Further Reading

- For a review on recent advances in the field of tetracycline antibiotics see: V. J. Lee, *Expert Opin. Ther. Pat.,* **1995**, *5*, 787.
- For a review on the mechanism of bacterial resistance to tetracycline antibiotics see: D. E. Taylor and A. Chau, *Antimicrob. Agents Chemother.*, **1996**, *40*, 1.

35. Synthesis of a D-*chiro*-Inositol 1-Phosphate

C. Jaramillo, J-L Chiara and M. Martín-Lomas *J. Org. Chem.*, **1994**, *59*, 3135.

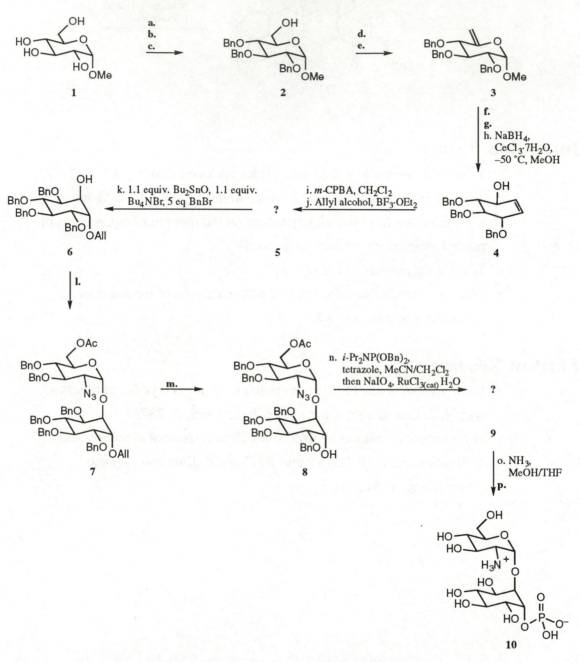

Discussion Points

- Give a mechanism for the reactions covered in steps **f** and **g**.
- Propose explanations for the selectivities obtained in steps **h**, **i** and **j**.
- Rationalise the regioselectivity of the benzylation carried out in step **k**.

36. Stereoselective Synthesis of (±)-Aromaticin

G. Majetich, J.-S. Song, A. J. Leigh and S. M. Condon, *J. Org. Chem.*, **1993**, *58*, 1030.

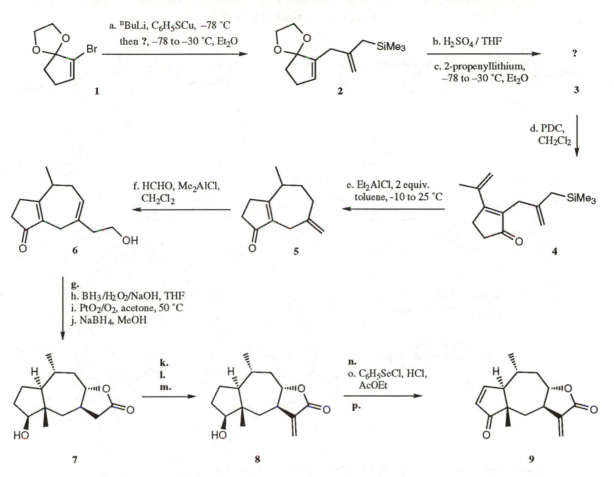

Discussion Points

- What is the mechanism of step **d**?
- Give reasons for the use of two equivalents of Lewis acid in step **e**.
- During the ene reaction of step **f**, a regioisomer of **6** could, in theory, be formed. Propose a structure for this isomer and suggest a reason why its formation was not observed.
- Explain the stereoselectivity in steps **h** and **j**.
- Suggest a possible reason why the phenylselenation reaction of step **o** required acid catalysis.

Further Reading

- For a review on the use of silanes in the synthesis of natural products, see: E. Langkopf and D. Schinzer, *Chem. Rev.*, **1995**, *95*, 1375.

37. Synthesis of 7-Methoxycyclopropamitosene

A. S. Cotterill, P. Hartopp, G. B. Jones, C. J. Moody, C. L. Norton, N. O'Sullivan and E. Swann,

Tetrahedron, **1994**, *50*, 7657.

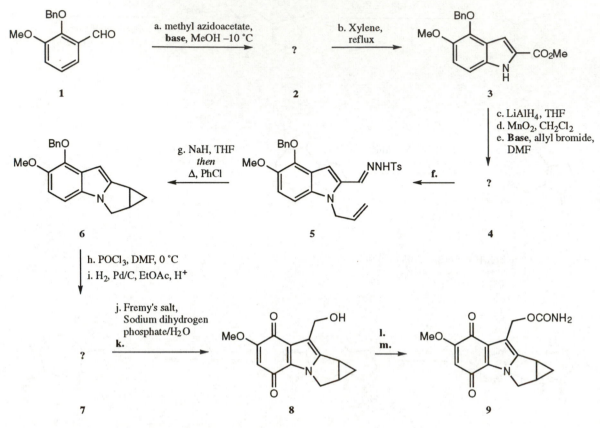

Discussion Points

- Suggest bases suitable for employment in steps **a** and **e**.
- Give a mechanism for the indole ring formation in step **b**.
- In step **b**, along with the desired compound **3**, a crystalline solid **10** was also obtained in 10–15% yield. An analytically pure sample provided the following data; C, 69.5; H, 5.4; N, 4.4%; *m/z* M$^+$ = 311. The nujol IR spectrum displayed key peaks at 1734 and 2234 cm^{-1}. The proton NMR spectrum gave, along with eight proton signals between 7.50 and 6.99 ppm, a 2-proton AB system (J = 12.3 Hz) centred at 5.21 ppm and two 3-proton singlets at 3.92 and 3.68, a singlet integrating for 1 proton at 5.04 ppm. Propose a possible structure for **10** indicating a mechanism for its formation.
- What is the mechanism of the cyclopropanation step **g**? How does this step differ in the number of equivalents of base required with respect to a Shapiro reaction?

38. Total Synthesis of (+)-γ-Lycorane

H. Yoshizaki, H. Satoh, Y. Sato, S. Nukui, M. Shibasaki and M. Mori, *J. Org. Chem.*, **1995**, *60*, 2016.

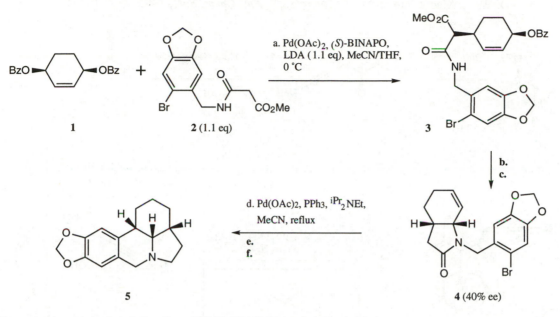

Discussion Points

- What is the structure of the palladium complex formed in reaction **a**? What is the rationale for enantioselectivity of the reaction?

- When NaH was used as the base in the palladium-catalysed alkylation step **a**, no enantioselectivity was observed. An enantioselective reaction was obtained with the use of a lithium base or alternatively via the transformation of **2** into its silylated derivative **6**. Propose an explanation for this finding.

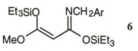

- What is the mechanism of step **d**?

Further Reading

- For a review on the selectivity in palladium-catalysed allylic substitutions, see: C. G. Frost, J. Howarth and J. M. J. Williams, *Tetrahedron: Asymmetry*, **1992**, *3*, 1089.

- For reviews on the Heck reaction, see: W. Cabri and I. Candiani, *Acc. Chem. Res.*, **1995**, *28*, 2; K. Ritter, *Synthesis*, **1993**, *8*, 735.

- For a review on enantioselective organotransition metal mediated reactions, see: S. L. Blystone, *Chem. Rev.*, **1989**, *89*, 1663.

39. Enantioselective Total Synthesis of (–)-7-Deacetoxyalcyonin Acetate

D. W. C. MacMillan and L. E. Overman, *J. Am. Chem. Soc.*, **1995**, *117*, 10391.

Discussion Points

- Suggest a structure for compound **4** and a possible way of synthesising it from **14**.

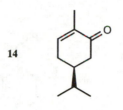

- Explain the stereoselectivity observed in step **e**.
- Suggest a mechanism for the formation of compound **7**.
- Which enantiomer of diethyl tartrate is necessary to use in order to obtain compound **9**?
- Give an explanation of the regioselectivity obtained in the epoxide reduction of step **k**.

Further Reading

- For another application of the Prins-pinacol rearrangement, see: G. C. Hirst, T. O. Johnson and H. E. Overman, *J. Am. Chem. Soc.*, **1993**, *115*, 2992.
- For a recent review on the Sharpless epoxidation reaction, see: A. D. Gupta, D. Bhuniya and V. K. Singh, *J. Indian Inst. Sci.*, **1994**, *74*, 71.
- For a review on the Nozaki–Hiyama reaction, see: P. Cintas, *Synthesis*, **1992**, 248.
- For a review of the use of TPAP as a catalytic oxidant, see: S. V. Ley, J. Norman, W. P. Griffith and S. P. Marsden, *Synthesis*, **1994**, 639.

40. Total Synthesis of (+)-Tetrahydrocerulenin

M. Miller and L. S. Hegedus, *J. Org. Chem.*, **1993**, *58*, 6779.

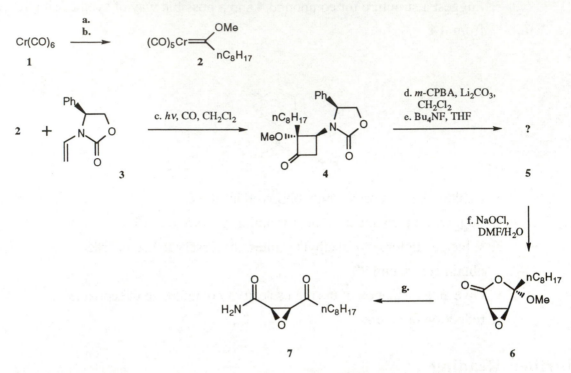

Discussion Points

- What is the mechanism of step **c**?
- Give reasons for the diastereoselectivity observed in step **c**.
- Suggest a synthesis of optically pure **3**.
- What is the mechanism of the Baeyer-Villiger reaction?

Further Reading

- For some references on the use of chromium carbenes, see: L. S. Hegedus, *Acc. Chem. Res.*, **1995**, *28*, 299; M. A. Schwindt, J. R. Miller, and L. S. Hegedus, *J. Organomet. Chem.*, **1991**, *413*, 143; L. S. Hegedus, *Pure Appl. Chem.*, **1990**, *62*, 691.

- For a review on the Baeyer-Villiger reaction, see: G. R. Krow, *Org. React.*, **1993**, *43*, 251. For the enzyme-catalysed version of this reaction, see: S. M. Roberts and A. J. Willetts, *Chirality*, **1993**, *5*, 334; M. C. Pirrung and R. S. Wilhelm, *Chemtracts: Org. Chem.*, **1989**, *2*, 29; C. T. Walsh and Y. C. J. Chen, *Angew. Chem.*, **1988**, *100*, 342.

- For the application of fluoride ion as a base in sugar chemistry, see: F. Santoyo-Gonzalez and F. Fernando-Mateo, *Synlett*, **1990**, *12*, 715.

41. Total Synthesis of (+)-Longifolene

B. Lei and A. G. Fallis, *J. Org. Chem.*, **1993**, *58*, 2186.

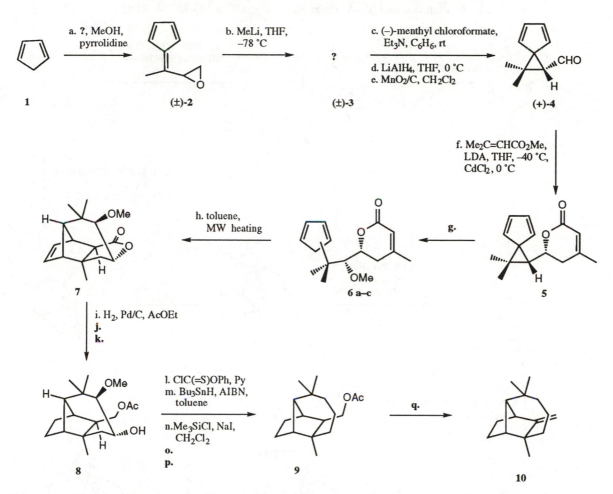

Discussion Points

- Explain the selectivity obtained in step **f** and give reasons for the use of cadmium chloride in the condensation reaction.
- Which one of the cyclopentadienyl regioisomers **6 a–c** leads to compound **7**? Propose structures for the other possible Diels–Alder adducts.

Further Reading

- For a review on the use of microwave heating, see: G. Bond, R. B. Moyes and D. A. Whan, *Cat. Today*, **1993**, *17*, 427.
- For a study of the influence of tether length on intramolecular Diels–Alder reactions, see: J. R. Stille and R. H. Grubbs, *J. Org. Chem.*, **1989**, *54*, 434.

42. Studies Toward the Total Synthesis of Cerorubenic Acid-III

L. A. Paquette and M.-A. Poupart, *J. Org. Chem.,* **1993**, *58*, 4245.

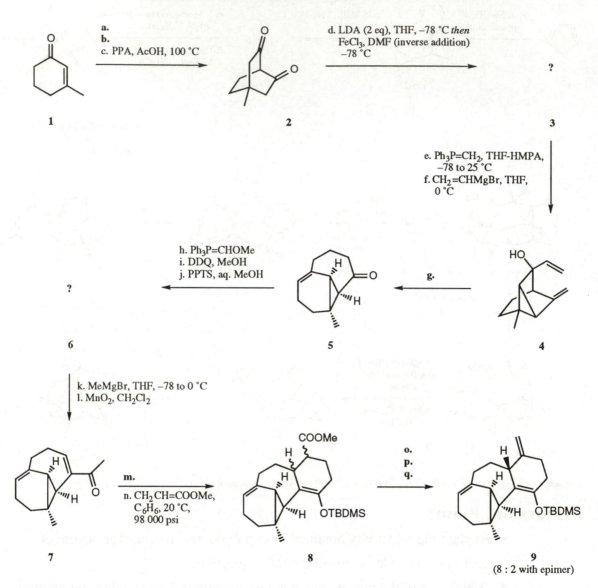

Discussion Points

- What is the purpose of FeCl₃ in step **d**?
- Give reasons for the high stereoselectivity observed in nucleophilic additions to the cyclopropane-containing system (e.g. step **f**).
- When the Wittig / Grignard addition sequence was carried out in the inverse order, compound **10** was isolated instead. How can this result be explained?

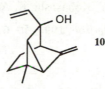

- Explain the stereoselectivity and the complete regioselectivity observed in the Diels–Alder reaction of step **n**.

Further Reading

- For further studies towards cerorubenic acid-III, see: L. A. Paquette, S. Hormuth and C. J. Lovely, *J. Org. Chem.*, **1995**, *60*, 4813; L. A. Paquette, G. Y. Lassalle and C. J. Lovely, *J. Org. Chem.*, **1993**, *58*, 4254; L. A. Paquette, D. N. Deaton, Y. Endo and M.-A. Poupart, *J. Org. Chem.*, **1993**, *58*, 4262.
- For a review on anion-assisted sigmatropic rearrangements, see: S. R. Wilson, *Org. React.*, **1993**, *43*, 93.
- For reviews on the oxy-Cope rearrangement, see: K. Durairaj, *Curr. Sci.*, **1994**, *66*, 917; L. A. Paquette, *Synlett.*, **1990**, *2*, 67.
- For another application of oxidative dianion coupling in natural product synthesis, see: J. L. Belletire and D. F. Fry, *J. Org. Chem.*, **1988**, *53*, 4724.
- For a review on the enhancement of rate and selectivity in Diels–Alder reactions, see: U. Pindur, G. Lutz and C. Ott, *Chem. Rev.*, **1993**, *93*, 741.

43. Total Synthesis of 10-Decarboxyquinocarcin

T. Katoh, M. Kirihara, Y. Nagata, Y. Kobayashi, K. Arai, J. Minami and S. Terashima,

Tetrahedron, **1994**, *50*, 6239.

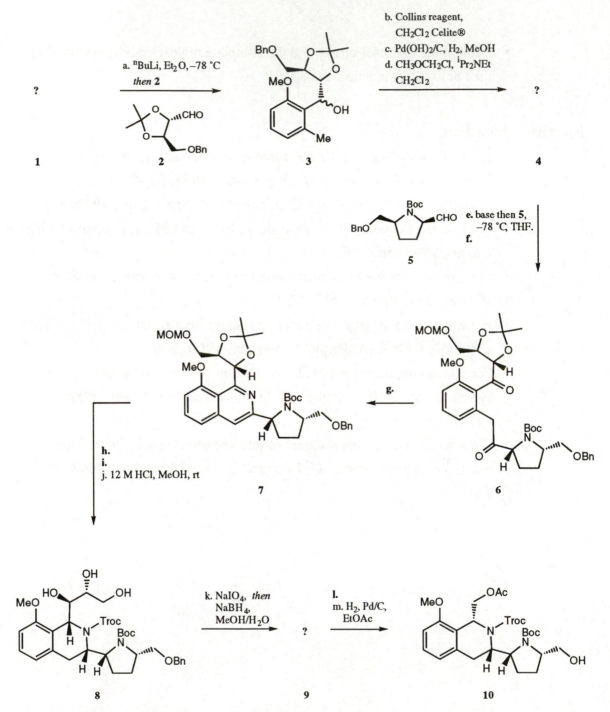

Abstracted with permission from *Tetrahedron*, **1994**, *50*, 6239 ©1994 Elsevier Science Ltd

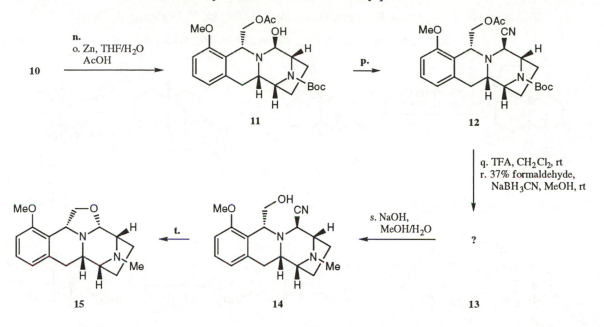

Discussion Points

- Suggest a suitable base for the alkylation of the 2-formylpyrrolidine **5**.
- Considering the overall strategy adopted, what could be considered the main motive for using the D-threose derivative **2** in the synthesis?
- Attempts at carrying out the sequence depicted in step **e** with the benzyl protecting group originally present on the chiral auxiliary **2** did not give the desired product. Give a possible motive for this observation.
- Propose a mechanism for the cyclisation step **g**. Give two other methods of synthesising an isoquinoline ring system.
- The reduction of the isoquinoline nucleus of compound **7** proceeded with high stereocontrol. What factors could be responsible for the selectivity?
- What is the mechanism involved in the deprotection of the Troc group?
- The cyclisation sequence from alcohol **10** gave a single hemiaminal **11**. Suggest an explanation for the stereochemical outcome of this reaction.
- Conversion of hemiaminal **11** into the nitrile **12** proceeds with retention of configuration. Propose a mechanism to rationalise this fact.

Further Reading

- For a review of directing groups controlling metallation reactions see: V. Snieckus, *Pure Appl. Chem.*, **1990**, *62*, 2047.
- For a a synthesis of the related system (±)-quinocarcinamide see: M. E. Flanagan and R. M. Williams, *J. Org. Chem.*, **1995**, *60*, 6791.

44. Total Synthesis of (+)-Pyripyropene A

T. Nagamitsu, T. Sunazuka, R. Obata, H. Tomoda, H. Tanaka, Y. Harigaya, S. Omura
and A. B. Smith III, *J. Org. Chem.*, 1995, *60*, 8126.

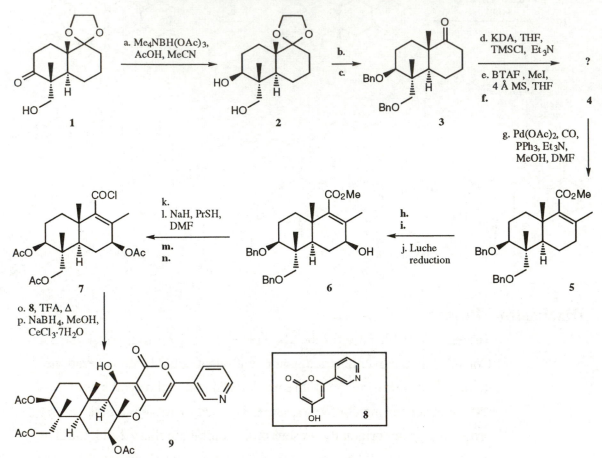

Discussion Points

- Explain the stereoselectivity observed in steps **a** and **p**.
- Give reasons for the use of the Kuwajima methylation protocol of steps **d** and **e**.
- What purpose does sodium thiolate serve in step **l**?
- Explain the mechanism of step **o**.

Further Reading

- For the biosynthesis of pyripyropene A, see: H. Tomoda, N. Tabata, Y. Nakata, H. Nishida, T. Kaneko, R. Obata, T. Sunazuka, and S. Omura, *J. Org. Chem.*, 1996, *61*, 882.
- For a review on the use of enol triflates in olefin synthesis, see: W. J. Scott and J. E. McMurry, *Acc. Chem. Res.*, 1988, *21*, 47.
- For the acylation reaction of 4-hydroxy-2-pyrone, see: E. Marcus, J. F. Stephen and J. K. Chan, *J. Heterocycl. Chem.*, 1969, *6*, 13.

45. Total Synthesis of *d,l*-Isospongiadiol

P. A. Zoretic, M. Wang, Y. Zhang and Z. Shen, *J. Org. Chem.*, **1996**, *61*, 1806.

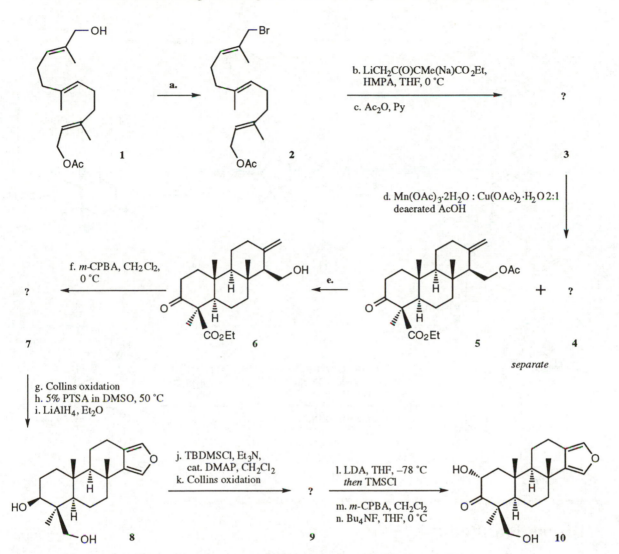

Discussion Points

- What kind of nOe would be expected to confirm the relative *trans* stereochemistry around the ring junctions in **5**?
- Suggest a mechanism for the formation of the furan ring upon treatment of the epoxy-aldehyde derived from **7** with acid.
- What is the mechanism of the Rubottom oxidation?

Further Reading

- For a review on manganese(III)-based oxidative free-radical cyclisations, see: B. M. Snider, *Chem. Rev.*, **1996**, *96*, 339.

46. Total Synthesis of the Stemona Alkaloid (–)-Stenine

P. Wipf, Y. Kim and D. M. Goldstein, *J. Am. Chem. Soc.*, **1995**, *117*, 11106.

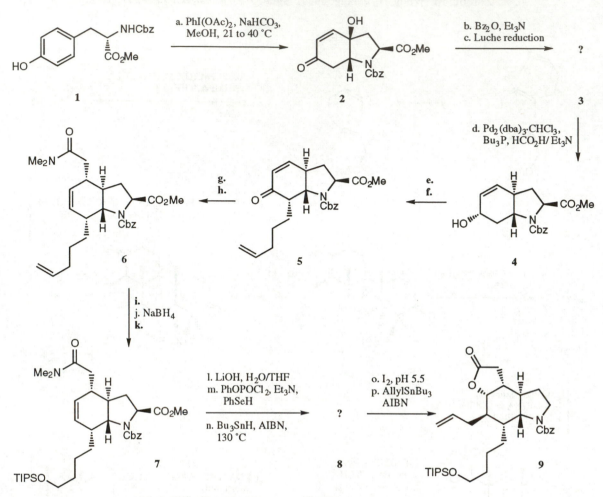

Abstracted with permission from *J. Am. Chem. Soc.*, **1995**, *117*, 11106 ©1995 American Chemical Society

Discussion Points

- Explain the diastereoselectivity in the formation of compound **2**.
- In the absence of base, substantial amounts of compounds **10** and **11** were isolated from reaction **d**. Give an explanation for their formation and for the stereoselectivity observed in the formation of **4**.

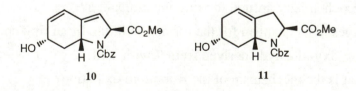

Further Reading

- For a related oxidative transformation of tyrosine derivatives, see: P. Wipf and Y. Kim, *J. Org. Chem.*, **1993**, *58*, 1649.
- For a similar approach in the conversion to the final compound, see: C.-Y. Chen and D. J. Hart, *J. Org. Chem.*, **1993**, *58*, 3840.

47. Total Synthesis of (–)-Papuamine

T. S. McDermott, A. A. Mortlock and C. H. Heathcock, *J. Org. Chem.*, **1996**, *61*, 700.

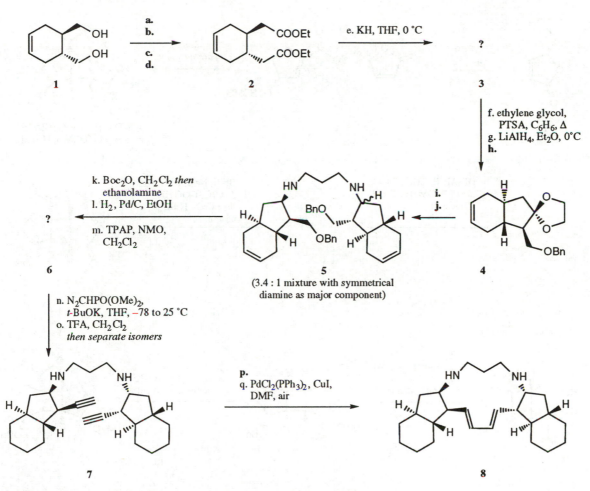

Abstracted with permission from *J. Org. Chem.*, **1996**, *61*, 700 ©1996 American Chemical Society

Discussion Points

- Suggest a synthesis of starting material **1**.
- What is the purpose of ethanolamine in the work-up of step **k**?
- What is the mechanism of step **n**?
- No coupling reaction occurred when step **q** was performed in the absence of CuI. Suggest a possible reason for this finding.

Further Reading

- For a related approach to the total synthesis of (+)-papuamine, see: A. G. M. Barrett, M. L. Boys and T. L. Boehm, *J. Org. Chem.*, **1996**, *61*, 685.
- For a review on the palladium-catalysed reaction of organotin compounds, see: T. N. Mitchell, *Synthesis*, **1992**, *9*, 803.

48. Total Synthesis of (+)-Stoechospermol

M. Tanaka, K. Tomioka and K. Koga, *Tetrahedron*, **1994**, *50*, 12829.

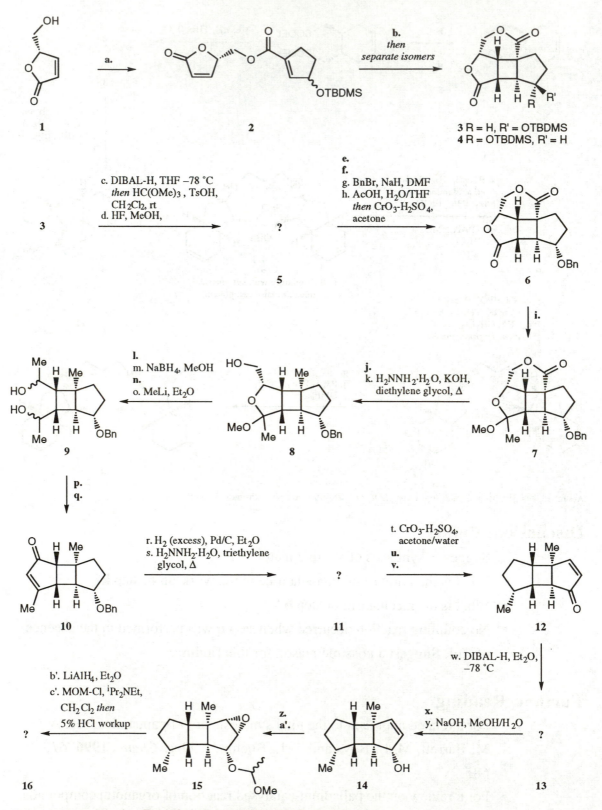

3 R = H, R' = OTBDMS
4 R = OTBDMS, R' = H

c. DIBAL-H, THF –78 °C
then HC(OMe)₃ , TsOH,
CH₂Cl₂, rt
d. HF, MeOH,

e.
f.
g. BnBr, NaH, DMF
h. AcOH, H₂O/THF
then CrO₃-H₂SO₄,
acetone

i.

l.
m. NaBH₄, MeOH
n.
o. MeLi, Et₂O

j.
k. H₂NNH₂·H₂O, KOH,
diethylene glycol, Δ

p.
q.

r. H₂ (excess), Pd/C, Et₂O
s. H₂NNH₂·H₂O, triethylene
glycol, Δ

t. CrO₃-H₂SO₄,
acetone/water
u.
v.

w. DIBAL-H, Et₂O,
–78 °C

b'. LiAlH₄, Et₂O
c'. MOM-Cl, ⁱPr₂NEt,
CH₂Cl₂ *then*
5% HCl workup

z.
a'.

x.
y. NaOH, MeOH/H₂O

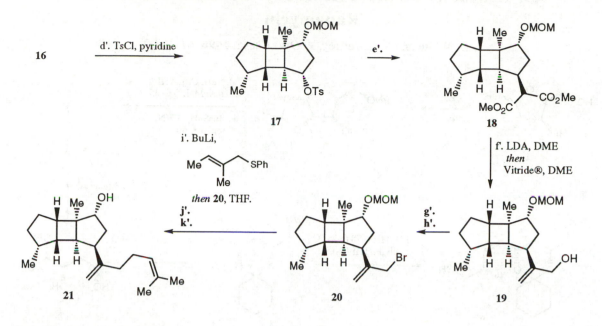

Discussion Points

- What is the mechanism of step **b**? Bearing in mind the fact that the 1:1 diastereomeric mixture **2** gave an approximately 1:1 mixture of isomers **3** and **4**, rationalise the stereoselectivity of the reaction and the formation of the four new chiral centres.

- Propose a reaction sequence which could convert the dilactone isomer **4** into compound **7**.

- What is the mechanism of the reduction with hydrazine hydrate in steps **k** and **s**?

- In step **w**, DIBAL-H is used to reduce an α,β-unsaturated ketone to an allylic alcohol. What reagent is usually used for such a transformation? Explain the stereoselectivity of this step.

- In the epoxidation of compound **14**, why is the α-epoxide the major isomer?

- In step **e´**, a tosylate is sucessfully displaced by an incoming malonate nucleophile with inversion of configuration. What is a potentially significant side-reaction when treating secondary tosylates with more basic nucleophiles?

- Suggest a mechanism for the reduction step **f´**.

Further Reading

- A racemic approach to stoechospermol has also been reported recently: M. Miesch, A. Cotte and M. Franck-Neumann, *Tetrahedron Lett.*, **1994**, *35*, 7031.

49. Synthesis of the Tricarbonyl Subunit C8–C19 of Rapamycin

J. D. White and S. C. Jeffrey, *J. Org. Chem.*, **1996**, *61*, 2600.

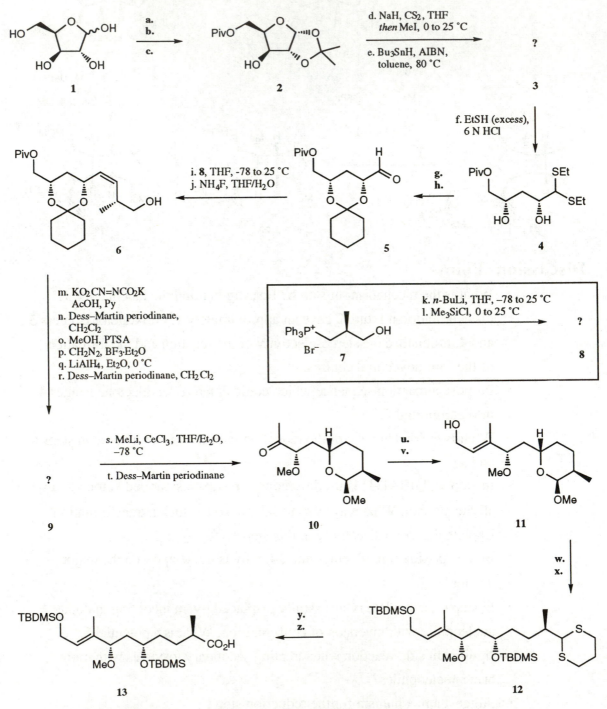

Abstracted with permission from *J. Org. Chem.*, **1996**, *61*, 2600 ©1996 American Chemical Society

49. Synthesis of the Tricarbonyl Subunit C8–C19 of Rapamycin

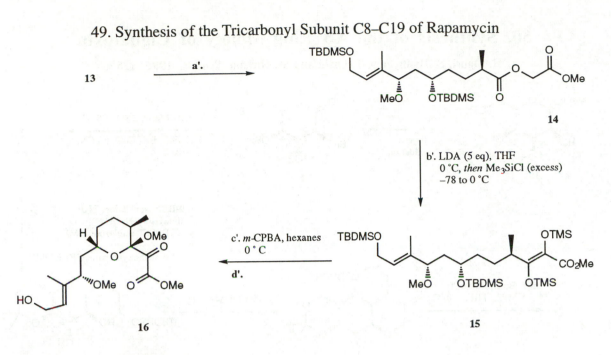

13 →ᵃ'→ **14**

b'. LDA (5 eq), THF
0 °C, *then* Me₃SiCl (excess)
−78 to 0 °C

c'. *m*-CPBA, hexanes
0 °C

d'.

16 ← **15**

Discussion Points

- Explain the selectivity observed in the Wittig olefination step **j** and suggest a reason why it was necessary to protect the hydroxyl group in **8**.
- What is the purpose of boron trifluoride etherate in step **p**?
- Explain the use of cerium trichloride in step **s**?
- Give a mechanism for the Chan rearrangement **b´**?
- What intermediate is presumably formed in the *m*-CPBA induced rearrangement of step **c´**?

Further Reading

- For studies on 1,2-anionic rearrangements in the gas phase, see: P. C. H. Eichinger, R. N. Hayes and J. H. Bowie, *J. Am. Chem. Soc.*, **1991**, *113*, 1949.
- For a review on the stereochemistry and mechanism of the Wittig reaction, see: E. Vedejs and M. J. Peterson, *Topics in Stereochemistry*, **1994**, *21*, 1.
- For studies on the Dess–Martin periodinane, see: D. B. Dess and J. C. Martin, *J. Am. Chem. Soc.*, **1991**, *113*, 7277; S. D. Meyer and S. L. Schreiber, *J. Org. Chem.*, **1994**, *59*, 7549.
- For some references on the Rubottom oxidation, see: J. Jauch, *Tetrahedron*, **1994**, *50*, 1203; R. Gleiter, M. Staib and U. Ackermann, *Liebigs Ann.*, **1995**, *9*, 1655.
- For studies on the presence of water in the CeCl₃/RLi system, see: W. J. Evans, J. D. Feldman and J. W. Ziller, *J. Am. Chem. Soc.*, **1996**, *118*, 4581.

50. Synthesis of the AB Ring Moiety of Ciguatoxin

H. Oguri, S. Hishiyama, T. Oishi and M. Hirama, *Synlett*, **1995**, 1252.

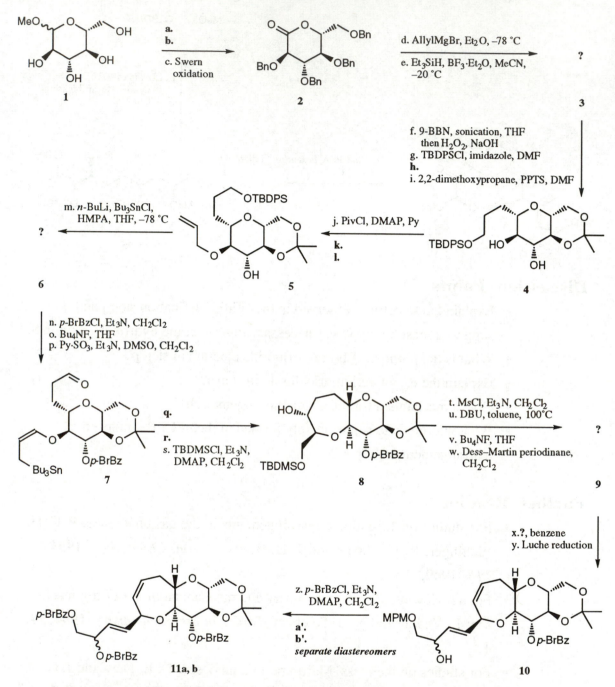

Discussion Points

- During the sequence **d–e** a new chiral centre is introduced. Which one of the steps is responsible for the stereoselectivity observed?

- In step **j** the selective protection of one hydroxyl group is accomplished. Suggest a reason for this selectivity.

- Give reasons for the stereoselectivity observed in the Yamamoto cyclisation of step **q**.

- Explain the regioselectivity of the DBU mediated elimination of step **u**.

- The compound obtained from Dess–Martin oxidation **w** proved to be unstable and had to be used without purification. Suggest a structure for the more stable compound formed on storage of **9**.

- Which analytical technique could be used to determine the absolute configuration of **11 a**, **b**?

Further Reading

- For reviews on sonochemistry, see: C. Einhorn, J. Einhorn and J.-L. Luche, *Synthesis*, **1989**, 787; J. L. Luche and C. Einhorn, *Janssen Chim. Acta*, **1990**, *8*, 8.

- For some references on the reaction of allylstannanes with aldehydes, see: H. Nakamura, N. Asao and Y. Yamamoto, *J. Chem. Soc., Chem. Commun.*, **1995**, *12*, 1273; J. Fujiwara, M. Watanabe and T. Sato, *J. Chem. Soc., Chem. Commun.*, **1994**, *3*, 349; V. Gevorgyan, I. Kadota and Y. Yamamoto, *Tetrahedron Lett.*, **1993**, *34*, 1313.

- For spectroscopic studies on the reaction, see: S. E. Denmark, E. J. Weber, T. M. Wilson and T. M. Willson, *Tetrahedron*, **1989**, *45*, 1053; G. E. Keck, M. B. Andrus and S. Castellino, *J. Am. Chem. Soc.*, **1989**, *111*, 8136.

- For some reviews on the use of circular dichroism, see: R. W. Woody, *Methods Enzymol.*, **1995**, *246*, 34; L. A. Nafie, *J. Mol. Struct.*, **1995**, *347*, 83.

- For some references on the hydrosilylation reaction, see: M. Onaka, K. Higuchi, H. Nanami and Y. Izumi, *Bull. Chem. Soc. Jpn.*, **1993**, *66*, 2638; M. Fujita and T. Hiyama, *J. Org. Chem.*, **1988**, *53*, 5405; Y. Nagai, *Org. Prep. Proceed. Int.*, **1980**, *12*, 13.

51. Total Synthesis of (+)-Carbonolide B

G. E. Keck, A. Palani and S. F. McHardy, *J. Org. Chem.*, **1994**, *59*, 3113.

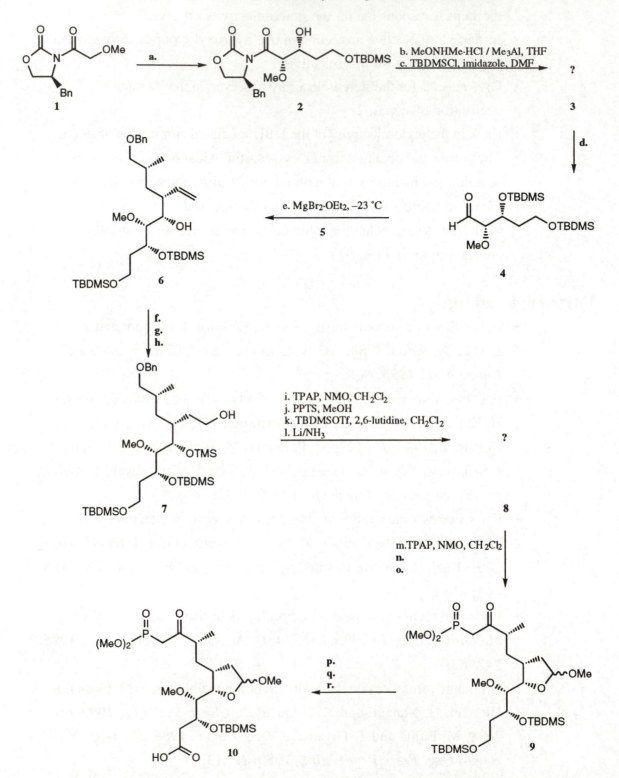

51. Total Synthesis of (+)-Carbonolide B

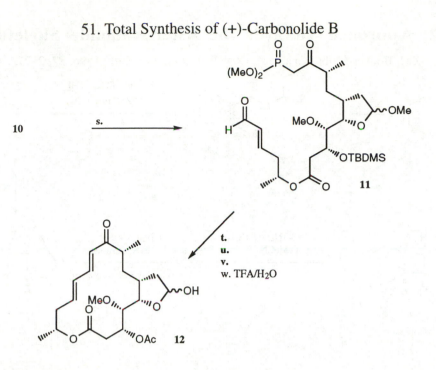

10 — s. → 11

12

Discussion Points

- The workup of step **b** includes washing with a saturated solution of Rochelle's salt. What is the purpose of this process?
- Suggest a structure for compound **5**.
- Rationalise the stereochemistry of the nucleophilic attack on compound **4** in step **e**. What role does the magnesium bromide etherate play in the stereocontrol?
- Propose a short synthesis of the alcohol used to esterify compound **10** starting from commercially available (R)-ethyl β-hydroxybutyrate.

Further Reading

- For a review of macrocyclic ring formation methods, see: Q. C. Meng and M. Hesse, *Top. Curr. Chem.*, **1991**, *161*, 107.
- For a review of the use of tetrapropylammonium perruthenate (TPAP) see: S. V. Ley, J. Norman, W. P. Griffith and S. P. Marsden, *Synthesis*, **1994**, 639.
- For a review, in English, of computer-aided conformational design in macrolide synthesis (including carbonolides) see: O. Yonemitsu, *Yuki Gosei Kagaku Kyokaishi*, **1994**, *52*, 946.
- The use of ultrasound to accelerate the hydroboration of double bonds has also recently been reported by H. Oguri, S. Hishiyama, T. Oishi and M. Hirama, *Synlett*, **1994**, 1252.
- For a more general review on the application of ultrasound to synthesis see: C. Einhorn, J. Einhorn and J.-L. Luche, *Synthesis*, **1989**, 787.

52. Approach Towards the Asteriscanolide Skeleton

K. I. Booker-Milburn and J. K. Cowell, *Tetrahedron Lett.*, **1996**, *37*, 2177.

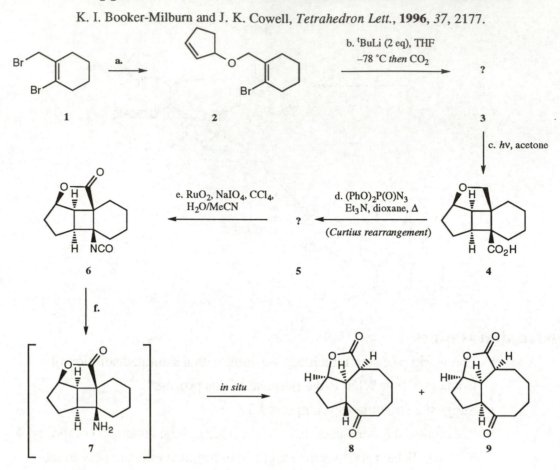

Abstracted with permission from *Tetrahedron Lett.*, **1996**, *37*, 2177 ©1996 Elsevier Science Ltd

Discussion Points

- What is the mechanism of the photolysis step **c**? What dictates the relative stereochemistry of the product **4**?

- Propose a mechanism for the rearrangement sequence leading to the epimeric cyclooctanones **8** and **9**.

- An attempt at oxidising the carboxylic acid **4** under the conditions reported in step **e** resulted in the formation of lactone **10**. Rationalise the formation of this product bearing in mind the mechanism of this kind of oxidation.

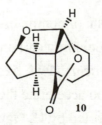

Further Reading

- For a similar photolytic approach to (+)-stoechospermol see: M. Tanaka, K. Tomioka and K. Koga, *Tetrahedron*, **1994**, *50*, 12829 (Exercise 48).

53. Total Synthesis of (+)-Dolabellatrienone

E. J. Corey and R. S. Kania, *J. Am. Chem. Soc.*, **1996**, *118*, 1229.

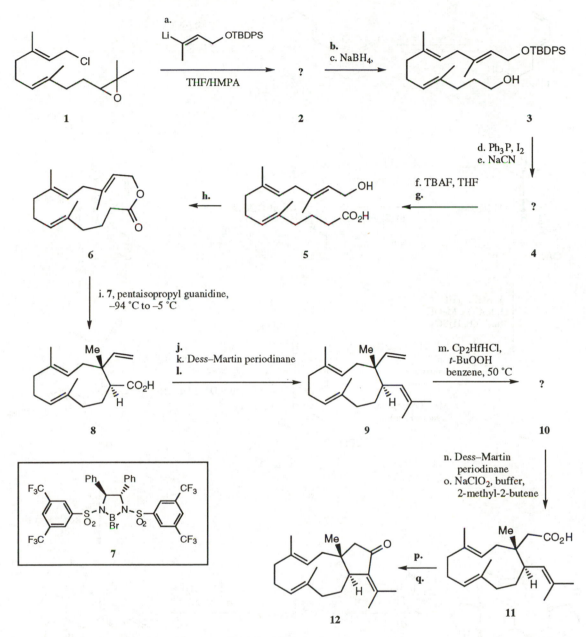

Discussion Points

- Suggest a synthesis of chloride **1** from (*E,E*)-farnesol?
- What is the mechanism of step **i**? Rationalise the stereochemical outcome in terms of the stereochemistry of the starting material **6** and the intermediates formed during the transformation.
- What symmetry does catalyst **7** possess?
- What reagents would usually be employed to effect the transformation of olefin **9** into **10** ?

54. Total Synthesis of Stenine

C.-Y. Chen and D. J. Hart, *J. Org. Chem.*, **1993**, *58*, 3840.

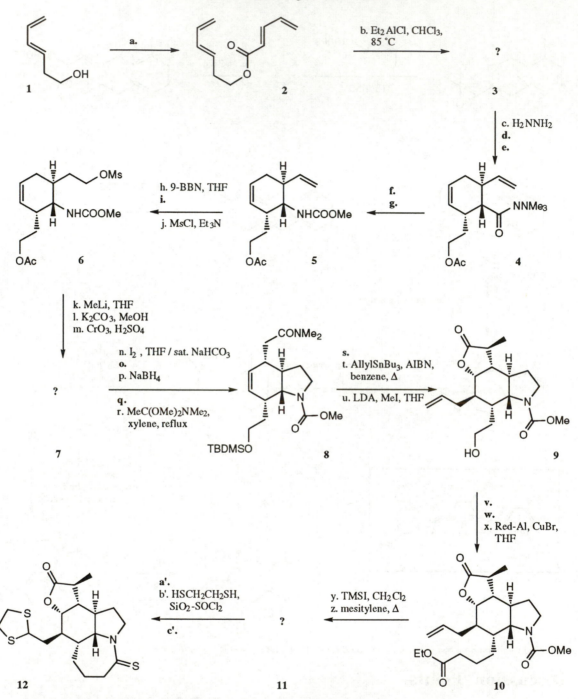

Abstracted with permission from *J. Org. Chem.*, **1993**, *58*, 3840 ©1993 American Chemical Society

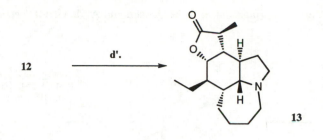

12 ——d'.——→ 13

Discussion Points

- Propose a structure for the other potential diastereoisomer arising from the intramolecular Diels–Alder cycloaddition.
- What is the mechanism of the Hoffman-type degradation of step **f**?
- A substantial amount of compound **14** is recovered if the Claisen rearrangement is attempted without selective protection of the primary hydroxyl group formed in step **p**. Explain the formation of this compound.

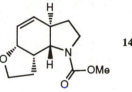

14

- How many equivalents of LDA are necessary in step **u**?

Further Reading

- For a review on diastereoselective Claisen rearrangements, see: H. J. Altenbach, *Org. Synth. Highlights*, **1991**, 115.
- For a study of Diels–Alder reaction on trienic esters in the absence and presence of a Lewis acid catalyst, see: M. Toyota, Y. Wada and K. Fukumoto, *Heterocycles*, **1993**, *35*, 111.
- For a recent NMR study of Lewis acid stoichiometry in Diels–Alder reactions, see: I. R. Hunt, C. Rogers, S. Woo, A. Rauk and B. A. Keay, *J. Am. Chem. Soc.*, **1995**, *117*, 1049; I. A. Hunt, A. Rauk and B. A. Keay, *J. Org. Chem.*, **1996**, *61*, 751.
- For a recent article on iodolactonisation leading to 7- to 12-membered ring lactones, see: B. Simonot and G. Rousseau, *J. Org. Chem.*, **1994**, *59*, 5912.

55. Synthesis of 3β-Acetoxydrimenin

H. J. Swarts, A. A. Verstegen-Haaksma, B. J. M. Jansen and A. de Groot, *Tetrahedron*, **1994**, *50*, 10083.

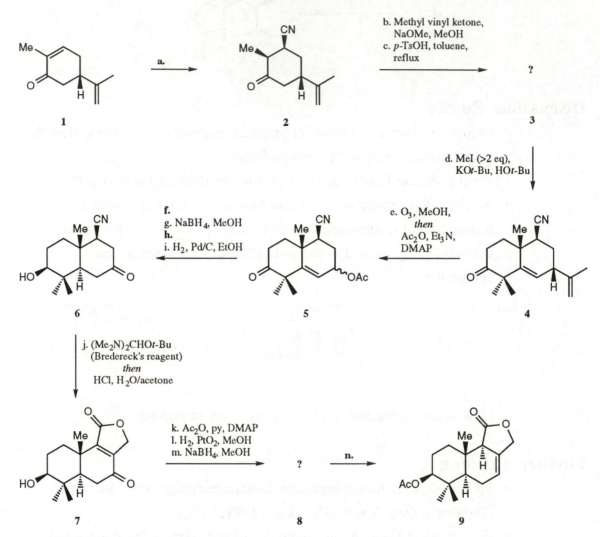

Discussion Points

- What governs the stereoselectivity of the 1,4-addition step **a**?
- Suggest a mechanism for the rearrangement in step **e**?
- Propose a mechanism for the non-basic formylation step **j**?
- Suggest a structure for a side-product of step **n** caused by the basicity of the DMAP.

Further Reading

- For the application of Criegee's rearrangement in the synthesis of enol ethers see: R. M. Goodman and Y. Kishi, *J. Org. Chem.*, **1994**, *59*, 5125.

56. The Total Synthesis of (+)-Adrenosterone

C. D. Dzierba, K. S. Zandi, T. Möllers and K. J. Shea, *J. Am. Chem. Soc.*, **1996**, *118*, 4711.

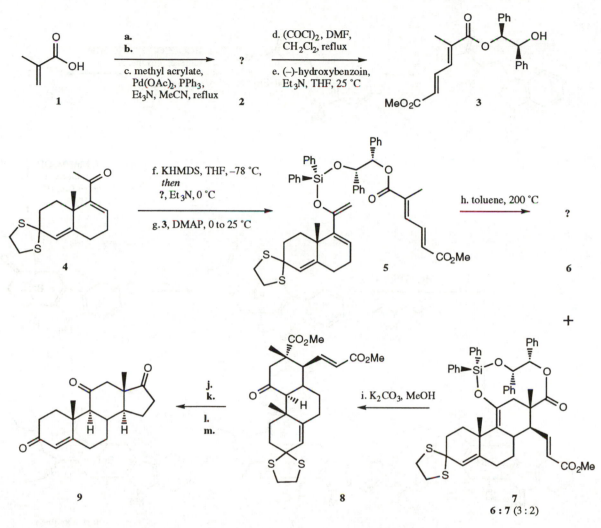

Abstracted with permission from *J. Am. Chem. Soc.*, **1996**, *118*, 4711 ©1996 American Chemical Society

Discussion Points

- When ethylene glycol was used instead of (–)-hydroxybenzoin (step **e**) in the preparation of the silicon tether, the intramolecular Diels–Alder reaction of step **h** proceeded with reverse diastereoselectivity, affording a mixture corresponding to **6:7** in a 1:10 ratio. Give an explanation for this finding.

Further Reading

- For a review on silicon-tethered reactions, see: M. Bols and T. Skrydstrup, *Chem. Rev.*, **1995**, *95*, 1253.

57. Total Synthesis of (−)-Suaveoline

X. Fu and J. M. Cook, *J. Org. Chem.*, **1993**, *58*, 661.

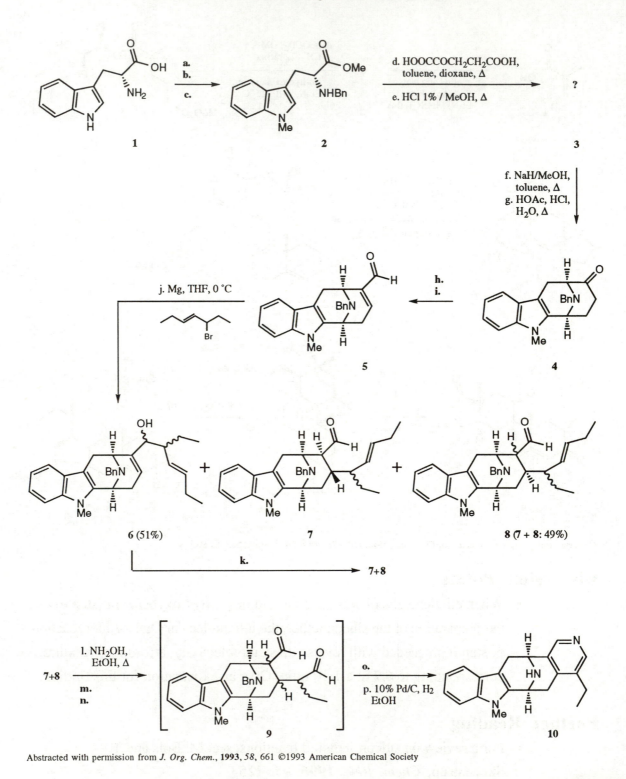

Discussion Points

- What is the mechanism of step **d**?
- Which of the products **6,7** and **8** is formally derived from a magnesium-ene reaction?
- Suggest a reason why it was necessary to protect the aldehydes **7** and **8** before the oxidative demolition steps **m** and **n**.
- When methanol was used as the solvent for debenzylation step **p**, compound **11** was isolated. Give an explanation for its formation.

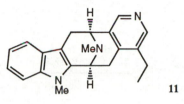

Further Reading

- For reviews on the Pictet–Spengler condensation and isoquinoline alkaloid synthesis, see: E. D. Cox and J. M. Cook, *Chem. Rev.*, **1995**, *95*, 1797; M. D. Rozwadowska, *Heterocycles*, **1994**, *39*, 903.
- For a review on the palladium-catalysed synthesis of condensed heteroaromatic compounds, see: T. Sakamoto, Y. Kondo and H. Yamanaka, *Heterocycles*, **1988**, *27*, 2225.
- For the use of amino acid esters in the synthesis of nitrogen heterocycles, see: H. Waldmann, *Synlett*, **1995**, *2*, 133.
- For the enantiospecific synthesis of related molecules, see: Y. Bi, L.-H. Zhang, L. K. Hamaker and J. M. Cook, *J. Am. Chem. Soc.*, **1994**, *116*, 9027.

58. Total Synthesis of (+)-7-Deoxypancratistatin

G. E. Keck, S. F. McHardy and J. A. Murry, *J. Am. Chem. Soc.*, **1995**, *117*, 7289.

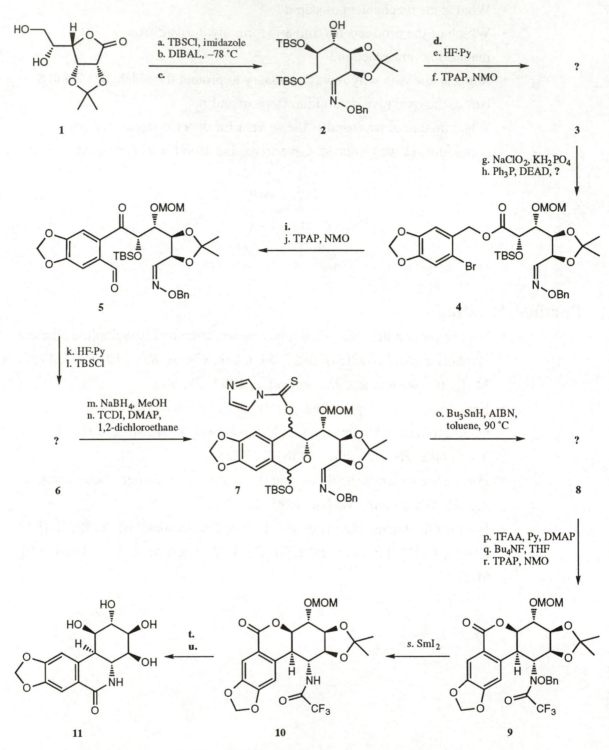

Discussion Points

- What is the mechanism of step **i**?
- The stereochemical outcome of the radical cyclisation step **o** is highly dependent on the use of a cyclic precursor (e.g. **7**). When acyclic compound **12** was submitted to the same reaction conditions as in **o**, it cyclised to give **13**. Give reasons for this finding.

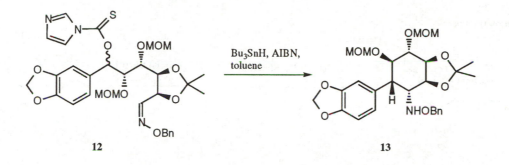

- Suggest alternative reagents for the reductive cleavage of the N–O bond (step **s**).

Further Reading

- For some reviews on the use of radical reactions in natural product synthesis, see: U. Koert, *Angew. Chem., Int. Ed. Engl.*, **1996**, *35*, 405; P. J. Parsons, C. S. Penkett and A. J. Shell, *Chem. Rev.*, **1996**, *96*, 195; D. P. Curran, J. Sisko, P. E. Yeske and H. Liu, *Pure Appl. Chem.*, **1993**, *65*, 1153.
- For the use of samarium iodide in organic synthesis, see: G. A. Molander, *Org. React.*, **1994**, *46*, 211; J. Inanaga, *Trends Org. Chem.*, **1990**, *1*, 23; J. A. Soderquist, *Aldrichimica Acta*, **1991**, *24*, 15.
- For some reviews on the Mitsunobu reaction, see: D. L. Hughes, *Org. Prep. Proced. Int.*, **1996**, *28*, 127; D. L. Hughes, *Org. React.*, **1992**, *42*, 335.

59. Enantioselective Synthesis of Deoxynojirimycin

A. J. Rudge, I. Collins, A. B. Holmes and R. Baker, *Angew. Chem., Int. Ed. Engl.*, **1994**, *33*, 2320.

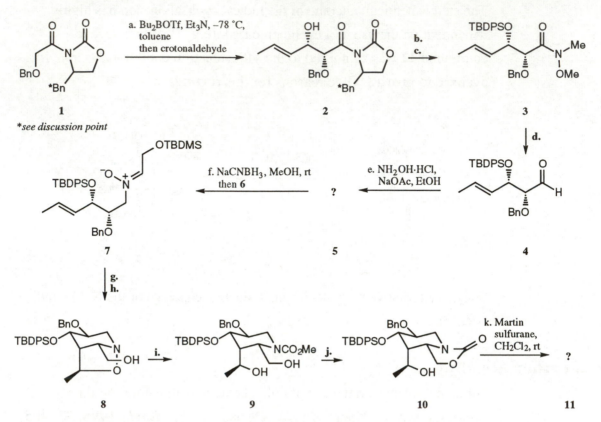

*see discussion point

Abstracted with permission from *Angew. Chem., Int. Ed. Engl.*, **1994**, *33*, 2320 ©1994 VCH Verlaggesellschaft

Discussion Points

- What should the absolute sterechemistry of the oxazolidinone benzyl group of compound **1** be in order to furnish the required product **2** under the conditions reported?
- Suggest a structure for compound **6**.
- Rationalise the stereoselectivity observed in step **g**.
- What is the mechanism of the Martin sulfurane step **k** leading to compound **11**?

Further Reading

- For a recent approach to nojirimycin and 1-deoxynojirimycin analogues see: N. Bentley, C. S. Dowdeswell and G. Singh, *Heterocycles*, **1995**, *41*, 2411; A. Kilonda, F. Compernolle and G. J. Hoornaert, *J. Org. Chem.*, **1995**, *60*, 5820.
- Martin's sulfurane was also recently used in the synthesis of (+)-lepicidin A: D. A. Evans and W. C. Black, *J. Am. Chem. Soc.*, **1993**, *115*, 4497.

60. Formal Synthesis of (±)-Aphidicolin

M. Toyota, Y. Nishikawa and K. Fukumoto, *Tetrahedron*, **1994**, *50*, 11153.

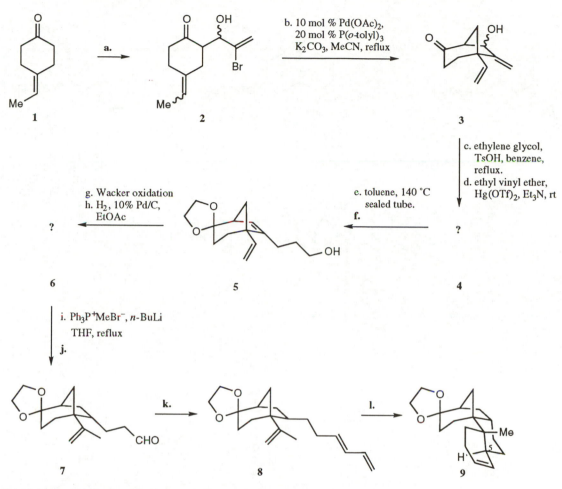

Abstracted with permission from *Tetrahedron*, **1994**, *50*, 11153 ©1994 Elsevier Science Ltd

Discussion Points

- The condensation step **a** gave a 3 : 1 mixture of isomers **2**. Assuming that the stereoselectivity of the reaction can be rationalised by a Zimmerman–Traxler transition state model, what should be the structure of the predominant isomer?

- What is the mechanism of step **b**?

- The tetracyclic product **9** was obtained in a 3 : 1 mixture with its epimer at C5. What is the mechanism of this reaction? Rationalise the stereochemical outcome.

Further Reading

- For a recent review of the Heck reaction see: W. Cabri and I. Candiani, *Acc. Chem. Res.*, **1995**, *28*, 2.

61. Synthetic Efforts Towards Bruceantin

C. K.-F. Chiu, S. V. Govindan and P. L. Fuchs *J. Org. Chem.*, **1994**, *59*, 311.

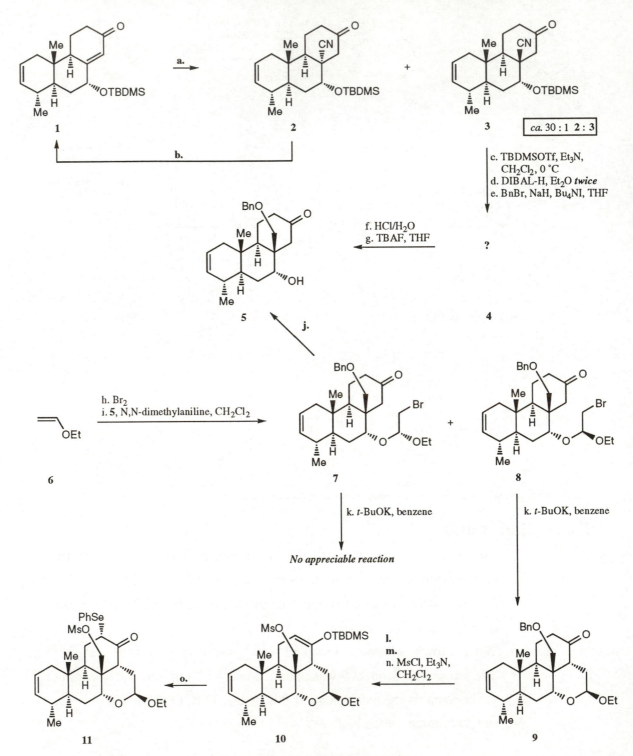

ca. 30 : 1 2 : 3

c. TBDMSOTf, Et₃N,
   CH₂Cl₂, 0 °C
d. DIBAL-H, Et₂O *twice*
e. BnBr, NaH, Bu₄NI, THF

f. HCl/H₂O
g. TBAF, THF

h. Br₂
i. **5**, N,N-dimethylaniline, CH₂Cl₂

k. *t*-BuOK, benzene

k. *t*-BuOK, benzene

No appreciable reaction

l.
m.
n. MsCl, Et₃N,
   CH₂Cl₂

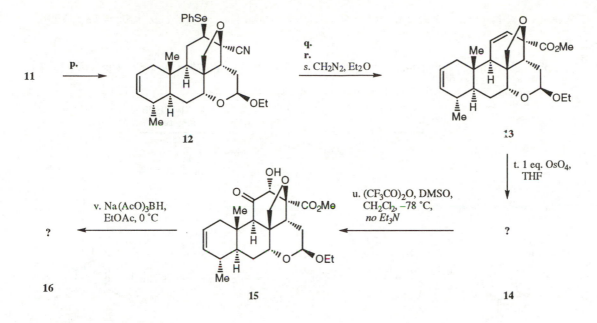

Discussion Points

- What is the controlling factor in the stereochemical outcome of the conjugate addition of cyanide to enone **1**?
- Based on the above fact, what would you predict the product of step **c** to be?
- Give a rationalisation for the difference in reactivity displayed by epimers **7** and **8** under basic conditions.
- In the transformation of ketone **11** into nitrile **12** the phenylselenyl group undergoes epimerisation. What is the driving force behind this process?
- Suggest a mechanism for the oxidation step **u** *carried out in the absence of triethylamine.*

Further Reading

- For a recent structure–activity relationship correlation of quassinoids as antitumor agents see: M. Okano, F. Narihiko, K. Tagahara, H. Tokuda, A. Iwashima, H. Nishino and K-H. Lee, *Cancer Lett.* **1995**, *94*, 139.
- For the application of trialkylamine/TMSOTf as a selective method for the formation of 'kinetic' silyl enol ethers of some α-aminocarbonyl cyclohexanones see: L. Rossi and A. Pecunioso, *Tetrahedron Lett.,* **1994**, *35*, 5285.

62. Total Synthesis of (+)-FR900482

T. Yoshino, Y. Nagata, E. Itoh, M. Hashimoto, T. Katoh and S. Terashima, *Tetrahedron Lett.*, **1996**, *37*, 3475.

T. Katoh, T. Yoshino, Y. Nagata, S. Nakatani and S. Terashima, *Tetrahedron Lett.*, **1996**, *37*, 3479.

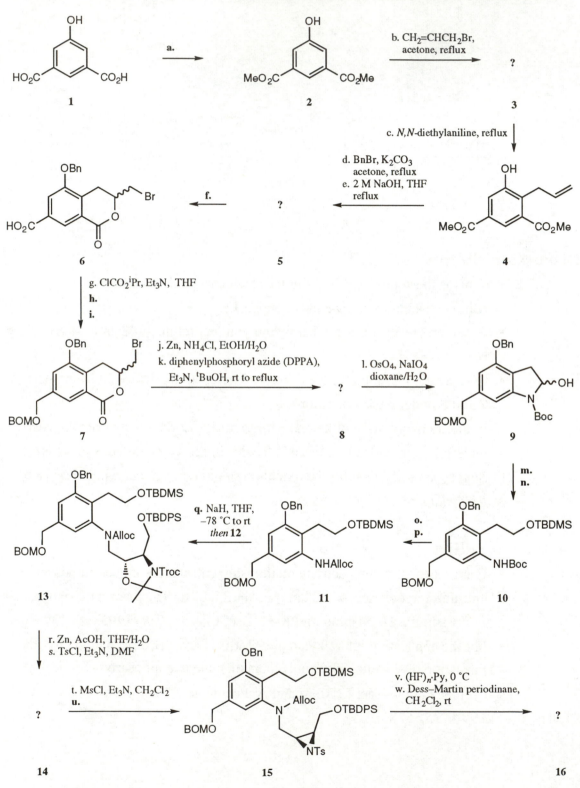

Abstracted with permission from *Tetrahedron Lett.*, **1996**, *37*, 3475 and 3479 ©1994 Elsevier Science Ltd

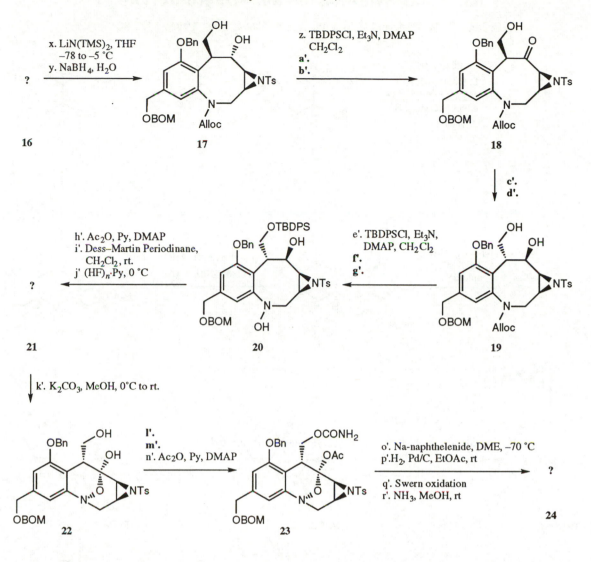

Discussion Points

- Apart from the high temperatures used in this synthesis, suggest alternative conditions which can be used to promote the transformation carried out in step **c**.

- What is the mechanism of step **k**?

- What are the standard conditions for the removal of a Boc protecting group from an amine? Are these conditions compatible with the deprotection of compound **10**?

- Propose a structure for compound **12**.

63. Total Synthesis of (3Z)-Dactomelyne

E. Lee, C. M. Park and J. S. Yun, *J. Am. Chem. Soc.*, **1995**, *117*, 8017.

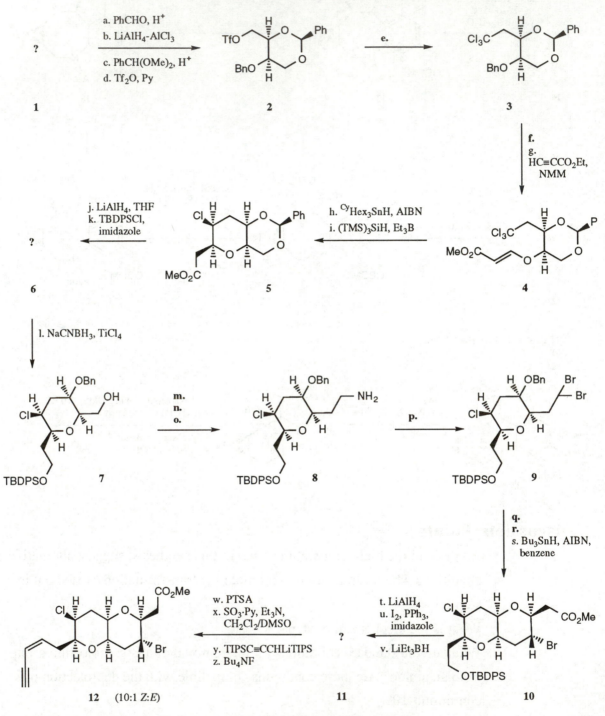

Discussion Points

- Explain the mechanism of step **b**.
- Explain the *cis*-2,6-selectivity observed in the radical cyclisation steps **h** and **s**.
- The *gem*-dichloro derivative obtained in step **h** could only be reduced stereoselectively with the use of a (trimethylsilyl)silane–triethylborane system. A variety of other systems only led to mixtures. Suggest a possible reason for this finding.
- Give a reason for the regioselectivity observed in the reductive cleavage of the benzylidene group in step **l**.
- In the radical cyclisation step **s**, none of the epimeric bromide was formed. When the reaction is applied to the synthesis of monocyclic products mixtures of epimers are usually obtained. Give an explanation for this difference in stereoselectivity.
- What is the mechanism of the Py·SO$_3$ oxidation?
- Explain the selective formation of the *Z* double bond in **12**.

Further Reading

- For some reviews on the use of radical reactions in natural product synthesis, see: U. Koert, *Angew. Chem. Int. Ed. Engl.*, **1996**, *35*, 405; P. J. Parsons, C. S. Penkett and A. J. Shell, *Chem. Rev.*, **1996**, *96*, 195; D. P. Curran, J. Sisko, P. E. Yeske and H. Liu, *Pure Appl. Chem.*, **1993**, *65*, 1153.
- For some reviews on the Peterson olefination reaction, see: A. G. M. Barrett, J. M. Hill, E. M. Wallace and J. A. Flygare, *Synlett*, **1991**, *11*, 764; D. J. Ager, *Org. React.*, **1990**, *38*, 1.
- For reviews on the use of different hydride donors in the Barton–McCombie deoxygenation, see: C. Chatgilialoglu and C. Ferreri, *Res. Chem. Intermed.*, **1993**, *19*, 755; S. David, *Chemtracts: Org. Chem.*, **1993**, *6*, 55.

64. Total Synthesis of (±)-Acerosolide

L. A. Paquette and P. C. Astles, *J. Org. Chem.*, **1993**, *58*, 165.

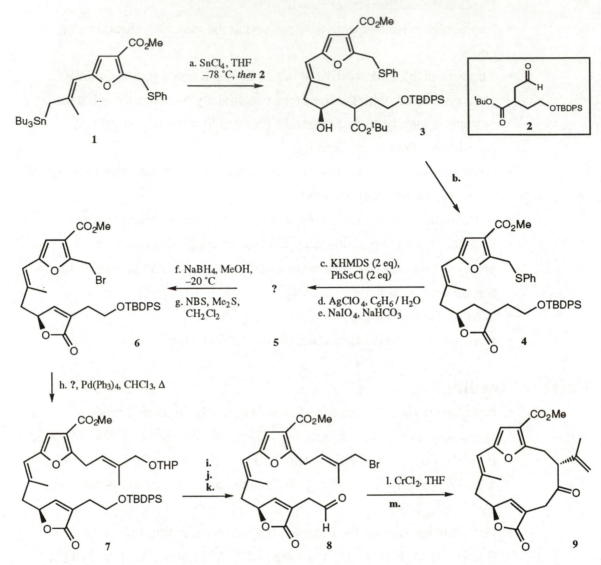

Abstracted with permission from *J. Org. Chem.*, **1993**, *58*, 165 ©1993 American Chemical Society

Discussion Points

- Treatment of aldehydes with allylstannane **1** in the presence of BF₃ would normally afford branched homoallylic alcohols. Suggest a mechanism for the regiochemical outcome of the reaction run in step **a**.
- What relative stereochemistry is to be expected from the Nozaki reaction of step **l**?

Further Reading

- For a review on the Nozaki reaction, see: P. Cintas, *Synthesis*, **1992**, 248.

65. Synthesis of a Vindorosine Precursor

J. D. Winkler, R. D. Scott and P. G. Williard, *J. Am. Chem. Soc.*, **1990**, *112*, 8971.

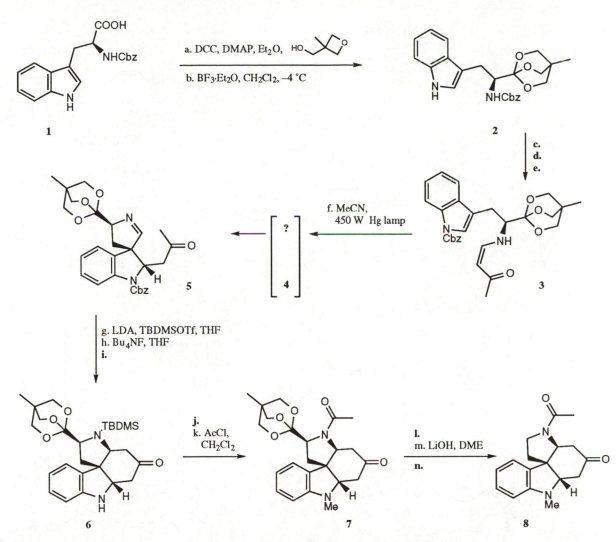

Discussion Points

- What is the mechanism of the protection sequence **a**, **b**?
- Explain the steroselectivities observed in the photochemical reaction of step **f** and in the Mannich reaction step **g**.
- What is the purpose of tetrabutylammonium fluoride in step **h**?
- What is the byproduct of step **k**?

Further Reading

- For reviews on tandem reaction in the synthesis of natural products, see P. J. Parsons, C. S. Penkett and A. J. Shell, *Chem. Rev.*, **1996**, *96*, 195; J. D. Winkler, C. M. Bowen and F. Liotta, *Chem. Rev.*, **1995**, *95*, 2003.

66. Synthesis of (–)-PGE₂ Methyl Ester

D. F. Taber and R. S. Hoerrner, *J. Org. Chem.*, **1992**, *57*, 441.

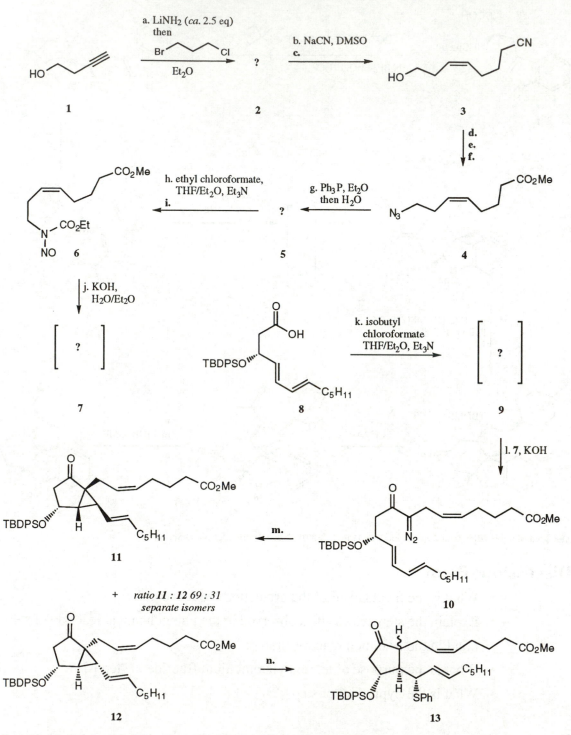

Abstracted with permission from *J. Org. Chem.*, **1992**, *57*, 441 ©1992 American Chemical Society

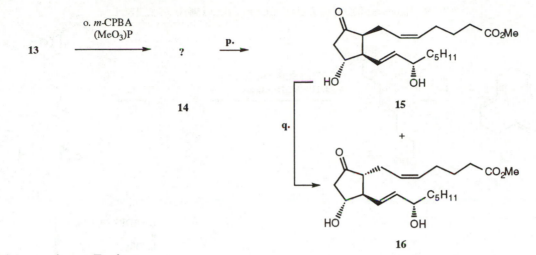

13 → o. *m*-CPBA (MeO₃)P → ? → p. → 14 ... 15 + 16

Discussion Points

- Propose a structure for the reactive intermediate formed in step **g** *before* the addition of water.
- Treating intermediate **7** with benzoic acid resulted in the formation of compound **17**. Suggest a structure for **7** and the mechanisms by which it gave both diazoketone **10** and benzoate ester **17**.

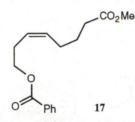

- What reactive intermediate is responsible for the formation of compounds **11** and **12**? Rationalise the stereoselectivity obtained.
- Suggest a mechanism for step **n**. What is the driving force behind this reaction?

Further Reading

- For a review on carbene insertion reactions see: D. J. Miller and C. J. Moody, *Tetrahedron*, **1995**, *51*, 10811.
- For an example of a palladium-catalysed annulation of a strained cyclopropane system see: R. C. Larock and E. K. Yum, *Tetrahedron*, **1996**, *52*, 2743.

67. Total Synthesis of (−)-Parviflorin

T. R. Hoye and Z. Ye, *J. Am. Chem. Soc.*, **1996**, *118*, 1801.

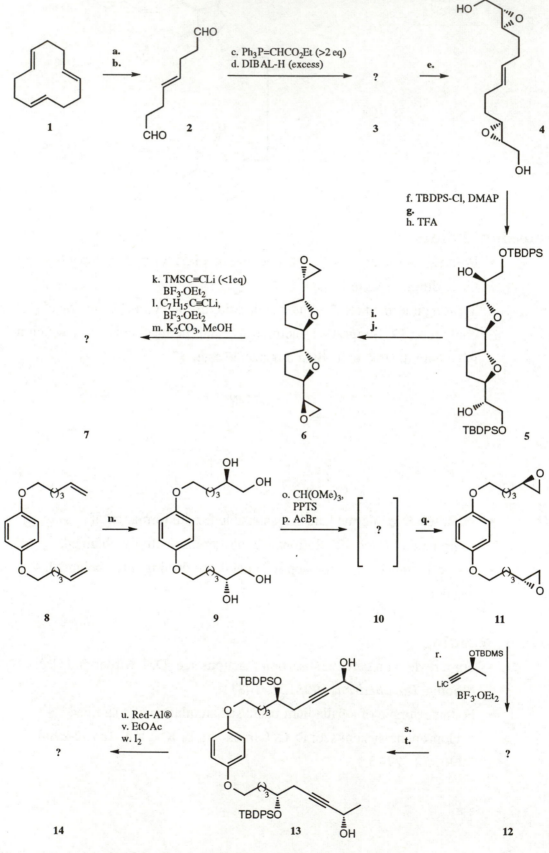

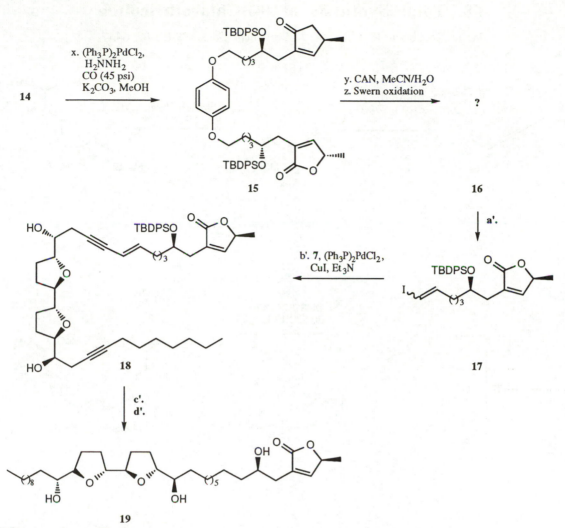

Discussion Points

- What kind of symmetry do the molecules from **4** to **6** and from **9** to **15** possess?

- Which isomer of diethyl tartrate would be expected to give the desired product in step **e**?

- In steps **i** and **j** the diol **5** is transformed into its corresponding bis-epoxide **6**. What product could reasonably have been expected had **5** been first desilylated, then treated as in steps **o** to **q**?

Further Reading

- For a review on two-directional chain synthesis see: C. S. Poss and S. L. Schreiber, *Acc. Chem. Res.*, **1994**, *27*, 9.

- For recent mechanistic investigations into the selectivity of the Sharpless asymmetric epoxidation see: P. G. Potvin and S. Bianchet, *J. Org. Chem.*, **1992**, *57*, 6629.

- For a review of palladium and copper mediated cross-coupling reactions see: R. Rossi, A. Carpita and F. Bellina, *Org. Prep. Proced. Int.*, **1995**, *27*, 127.

68. Total Synthesis of (–)-Chlorothricolide

W. R. Roush and R. J. Sciotti, *J. Am. Chem. Soc.*, **1994**, *116*, 6457.

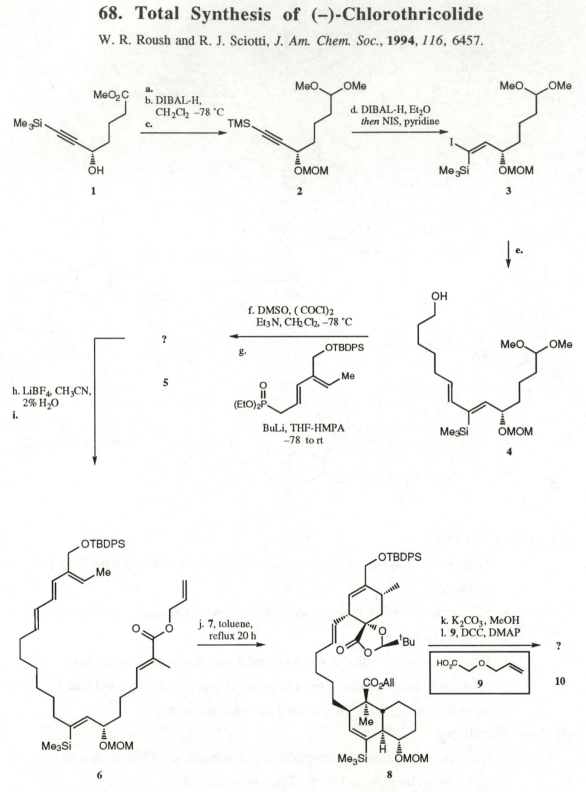

Abstracted with permission from *J. Am. Chem. Soc.*, **1994**, *116*, 6457 ©1994 American Chemical Society

68. Total Synthesis of (−)-Chlorothricolide

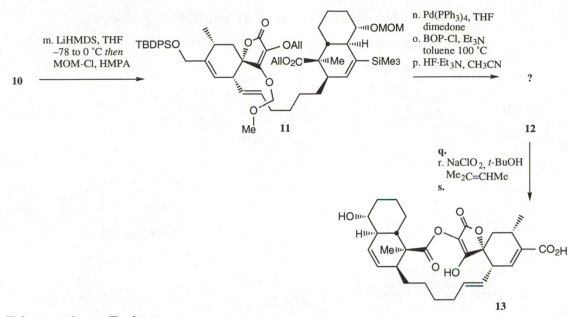

10 →(m. LiHMDS, THF −78 to 0 °C *then* MOM-Cl, HMPA)→ **11**

11 →(n. Pd(PPh₃)₄, THF dimedone / o. BOP-Cl, Et₃N toluene 100 °C / p. HF·Et₃N, CH₃CN)→ **12** ?

12 →(q. / r. NaClO₂, *t*-BuOH, Me₂C=CHMe / s.)→ **13**

Discussion Points

- Propose suitable methods for the synthesis of optically active alcohol **1** from its corresponding ketone.
- Suggest a mechanism for the iodination of compound **2**. What is the purpose of the DIBAL-H in the sequence?
- Give the mechanism of step **f**.
- Explain the stereoselectivity of steps **g** and **i**.
- What is the structure of **7** introduced in step **j**?
- Rationalise both the mechanism and stereocontrol that led to the cyclic system **8**, the major product of step **j**.
- What is the purpose of the 2-methyl-2-butene used in step **r**?

Further Reading

- For methods for enantioselective reduction of unsymmetrical ketones see: V. K. Singh, *Synthesis*, **1992**, 605.
- For an excellent review of intramolecular Diels–Alder reactions see: W. Carruthers, *Cycloaddition Reactions in Organic Synthesis,* Pergamon Press, Oxford, 1990.
- For a review of ring closure methods in the synthesis of natural products see: Q. Meng and M. Hesse, *Top. Curr. Chem.*, **1992**, *161,* 107.
- Suzuki cross-coupling reactions have been reviewed by N. Miyaura and A. Suzuki, *Chem. Rev.*, **1995**, 2457; A. Suzuki, *Pure Appl. Chem.*, **1994**, *66*, 213 and A. R. Martin and Y. Yang, *Acta Chem. Scand.*, **1993**, *47,* 221.
- A review of stereochemistry and mechanism in the Wittig reaction has been published by E. Vedejs and M. J. Peterson, *Top. Stereochem.*, **1994**, *21,* 1.

69. Synthetic Studies on Furanoheliangolides

D. S. Brown and L. A. Paquette, *J. Org. Chem.*, **1992**, *57*, 4512.

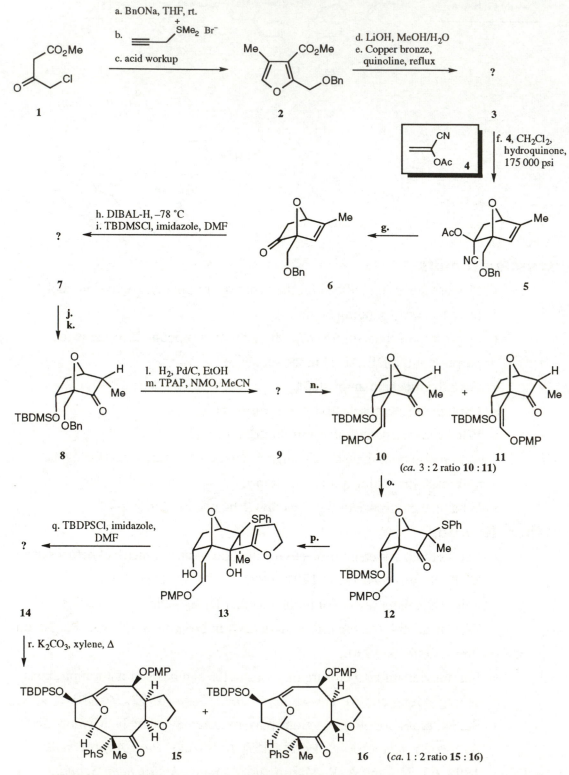

Abstracted with permission from *J. Org. Chem.*, **1992**, *57*, 4512 ©1992 American Chemical Society

Discussion Points

- Suggest a mechanism for the formation of furan **2**.
- Rationalise the stereochemical outcome of step **f**.
- The transformation **3** ⟶ **6** involves the formal [4+2] cycloaddition of a ketene to the furan ring. Why are ketenes not usually employed directly for Diels–Alder reactions?
- Propose a suitable mechanism for the rearrangement of compound **14** to give the isomers **15** and **16**. What role does the potassium carbonate play in the reaction?
- What product could be expected to be formed from the *E*-enol ether **17** under the conditions outlined in step **r**?

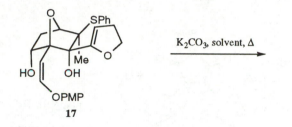

Further Reading

- For recent reviews of the oxy-Cope rearrangement see: K. Durairaj, *Curr. Sci.*, **1994**, *43*, 917; L. A. Paquette, *Synlett*, **1990**, 67. For a more general review on anion assisted sigmatropic rearrangements see: S. R. Wilson, *Org. React.*, **1993**, *43*, 93.
- For a review of the use of tetrapropylammonium perruthenate (TPAP) see: S. V. Ley, J. Norman, W. P. Griffith and S. P. Marsden, *Synthesis*, **1994**, 639.
- For some other recent applications of the Diels–Alder reaction of furans with ketene equivalents see: I. Yamamato and K. Narasaka, *Chem. Lett.*, **1995**, 1129; P. Metz, M. Fleischer and R. Frohlich, *Tetrahedron*, **1995**, *51*, 711 and K. Konno, S. Sagara, T. Hayashi and H. Takayama, *Heterocycles*, **1994**, *39*, 51.

70. Total Synthesis of Staurosporine

J. T. Link, S. Raghavan and S. J. Danishefsky, *J. Am. Chem. Soc.*, **1995**, *117*, 552.

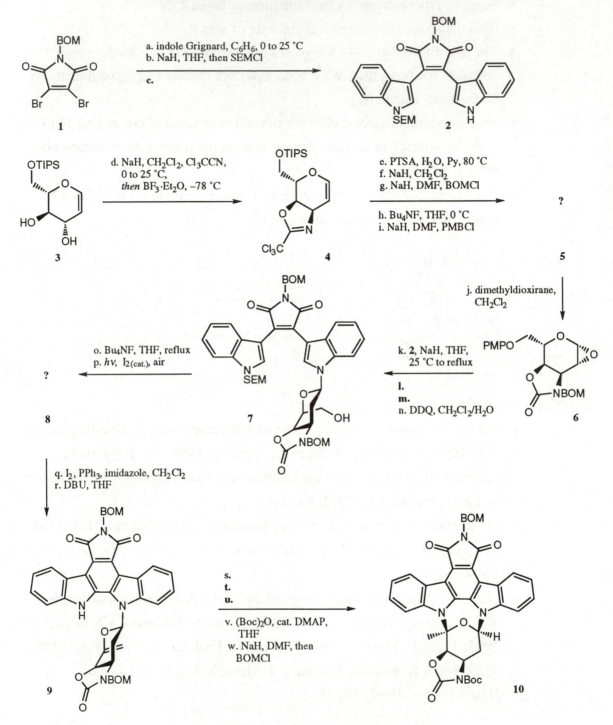

Abstracted with permission from *J. Am. Chem. Soc.*, **1995**, *117*, 552 ©1995 American Chemical Society

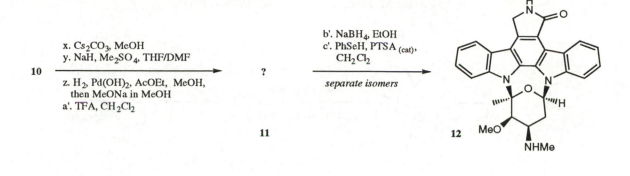

x. Cs$_2$CO$_3$, MeOH
y. NaH, Me$_2$SO$_4$, THF/DMF

10 →

z. H$_2$, Pd(OH)$_2$, AcOEt, MeOH,
   then MeONa in MeOH
a'. TFA, CH$_2$Cl$_2$

11

b'. NaBH$_4$, EtOH
c'. PhSeH, PTSA (cat),
   CH$_2$Cl$_2$

separate isomers

12

Discussion Points

- What is the structure of the intermediate formed in step **d**? Suggest a reason for the closure of the oxazoline ring on the ß-face (formation of **4**) and not on the α-face.

- Step **j** yielded a 2.5:1 mixture of α:ß epoxides. Give reasons for the low stereoselectivity observed.

- The minor epoxide was found to react less efficiently with **2**. Suggest a possible reason for this observation.

- The Boc protecting group on the oxazolidinone nitrogen of **10** plays a crucial role in steps **x** and **y**. Suggest the reason why it was introduced.

- What is the mechanism of step **c´**?

Further Reading

- For reviews on the use of dimethyldioxirane, see: G. Dyker, *J. Prakt. Chem./Chem. Ztg.*, **1995**, *337*, 162; W. Adam and L. Hadjiarapoglou, *Top. Curr. Chem.*, **1993**, *164*, 45.

- For examples of deoxygenation using benzeneselenol, see: S. Kuno, A. Otaka, N. Fujii, S. Funakoshi and H. Yajima, *Pept. Chem.*, **1986**, *23*, 143; M. J. Perkins, B. V. Smith, B. Terem and E. S. Turner, *J. Chem. Res., Synop.*, **1979**, *10*, 341.

71. Total Synthesis of (–)-Cephalotaxine

N. Isono and M. Mori, *J. Org. Chem.*, **1995**, *60*, 115.

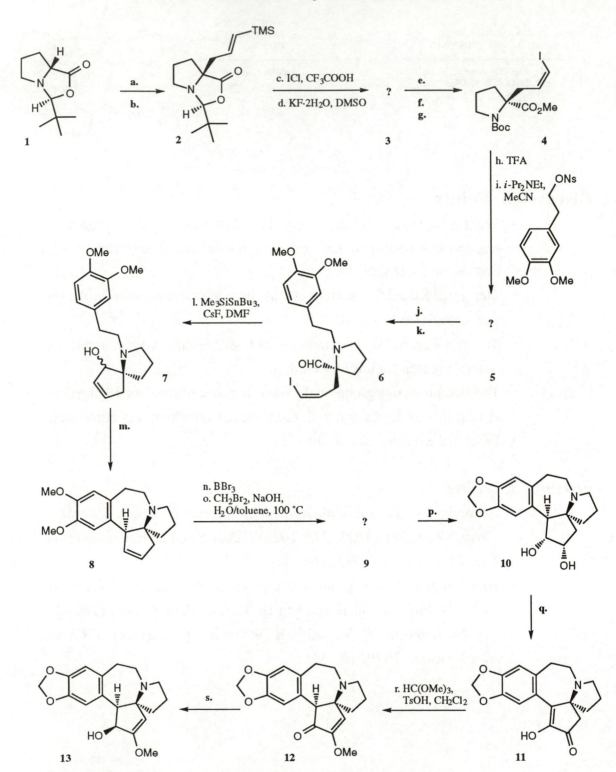

Discussion Points

- Give an explanation for the diastereoselectivity observed in the formation of compound **2**.
- A small amount of compound **14** is formed in step **l**. Explain the formation of this product.

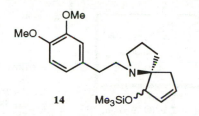

14

- What is the reactive species in step **l**?
- What is the mechanism for the formation of compound **8**?
- Give reasons for the stereoselectivity observed in the dihydroxylation step **p**.
- More drastic conditions (dimethoxyacetone, TsOH, dioxane, reflux) in the formation of methyl vinyl ether **12** led to racemic material. Propose a mechanism which accounts for the racemisation.

Further Reading

- For a review (in German) on asymmetric induction using proline derivatives, see: S. Blechert, *Nachr. Chem., Tech. Lab.*, **1979**, *27*, 768.
- For some reviews on asymmetric dihydroxylation, see: B. B. Lohray, *Tetrahedron: Asymmetry*, **1992**, *3*, 1317; H. C. Kolb, M. S. VanNieuwenhze and K. B. Sharpless, *Chem. Rev.*, **1994**, *94*, 2483.

72. The Synthesis of Picrotoxinin

B. M. Trost and M. J. Krische, *J. Am. Chem. Soc.*, **1996**, *118*, 233.

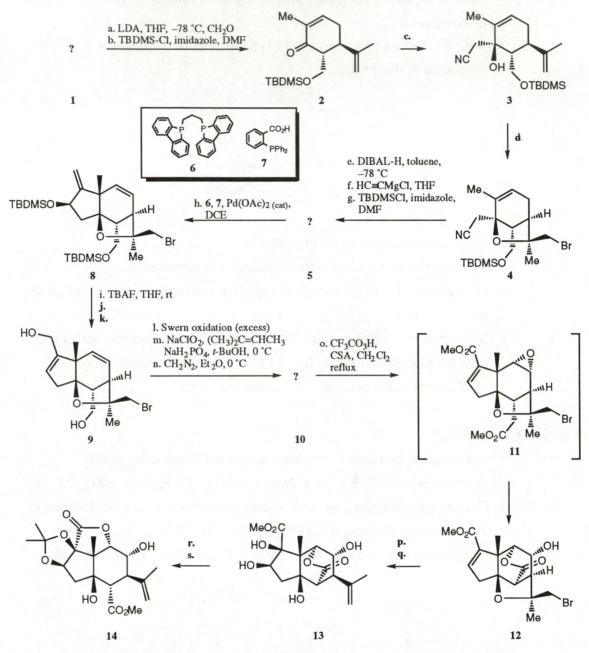

Abstracted with permission from *J. Am. Chem. Soc.*, **1996**, *118*, 233 ©1996 American Chemical Society

Discussion Points

- Propose a motive for the stereoselectivity of step **c**.
- Give a mechanism for step **h**.
- Bearing in mind the stereochemistry of epoxide **11**, propose a mechanism for its acid-catalysed ring opening under the conditions of step **o**.
- Suggest a possible reason for the increased stability of five-membered ring lactone **14** compared to the six-membered ring bridged system **13**.

73. Total Synthesis of (+)-Dactylol

G. Molander and P. R. Eastwood, *J. Org. Chem.*, **1995**, *60*, 4559.

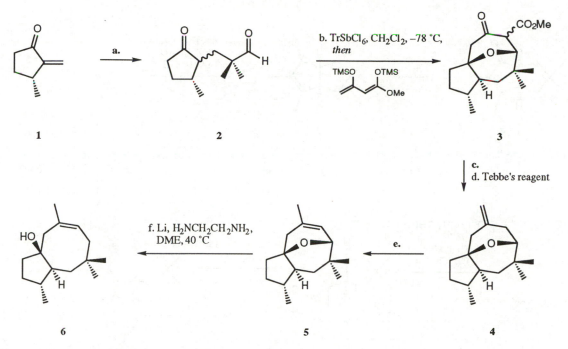

Discussion Points

- Propose a mechanism for the [3 + 5] annulation step **b**.
- What is the mechanism of the Tebbe methylenation reaction?
- The desired product **6** was isolated in only 25% yield. Bearing in mind the mechanism of the dissolving metal ring opening reaction, suggest a structure for the major product (36%) isolated from the reaction mixture.

Further Reading

- For other papers on [3+5] annulation reactions of bis(trimethylsilyl) enol ethers, see: G. A. Molander and P. R. Eastwood, *J. Org. Chem.*, **1996**, *61*, 1910; G. A. Molander and P. R. Eastwood, *J. Org. Chem.*, **1995**, *60*, 8382.
- For some reviews on the metal–ammonia reduction, see: P. W. Rabideau, *Tetrahedron*, **1989**, *45*, 1579; A. Rassat, *Pure Appl. Chem.*, **1977**, *49*, 1049.
- For a review on the methylenation reaction with Tebbe's reagent, see: H. U. Reissig, *Org. Synth. Highlights*, **1991**, 192.

74. Synthesis of (±)-Ceratopicanol

D. L. J. Clive, S. R. Magnuson, H. W. Manning and D. L. Mayhew,

J. Org. Chem.,**1996**, *61*, 2095.

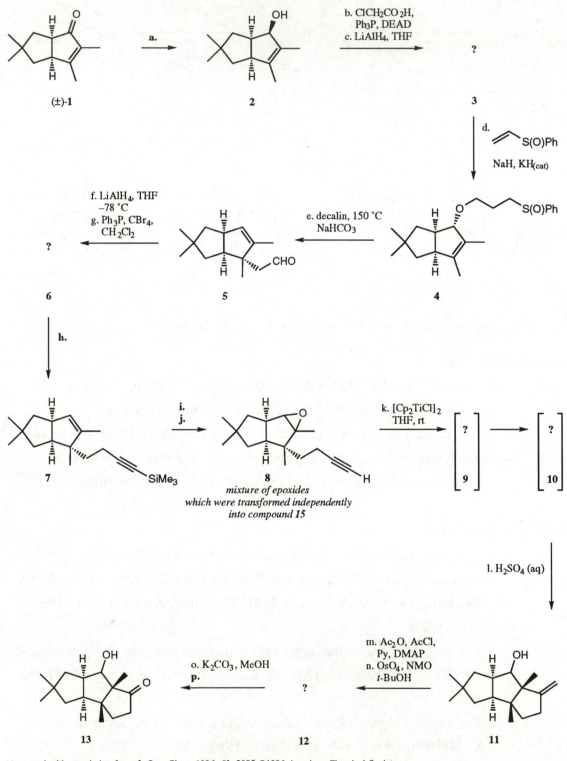

74. Synthesis of (±)-Ceratopicanol

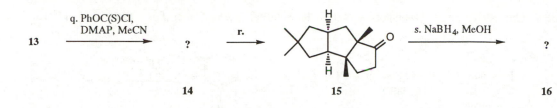

13 → [q. PhOC(S)Cl, DMAP, MeCN] → **14** ? → [r.] → **15** → [s. NaBH₄, MeOH] → **16** ?

Discussion Points

- What directs the selectivity of the reduction step **a**? What is the reagent usually employed for the reduction of less hindered α,β-unsaturated ketones?

- What is the mechanism of step **e**?

- In step **g** a Ph₃P–CBr₄ system is used after reducing the aldehyde **5**. What is the mechanism by which this system works?

- Had it been necessary to form a six-membered ring in step **k**, is it likely that the method used in this synthesis would have been sucessful?

- Step **k** proceeds via the ring-opening of the epoxide **8**. Suggest a reason for its regioselectivity, evident from the fact that the product formed is **11**.

- The overall transformation of compound **13** into **15** employs a classical deoxygenation protocol. Suggest other ways of deoxygenating alcohols or ketones.

Further Reading

- For reviews on radical cyclisations see: M. Malacria, *Chem. Rev.,* **1996**, *96*, 289; U. Koert, *Angew. Chem., Int. Ed. Engl.*, **1996**, *35*, 405; T. V. RajanBabu, *Acc. Chem. Res.*, **1991**, *24*, 139; D. P. Curran, *Synlett*, **1991**, 63; C. P. Jasperse, D. P. Curran and T. L Fevig, *Chem. Rev.*, **1991**, *91*, 1237.

- For a more general review on cationic, radical and anionic cyclisations see: C. Thebtaranotnth and Y. Thebtaranotnth, *Tetrahedron*, **1990**, *46*, 1385.

- For a review of the use of several classes of tandem reactions in organic synthesis see: P. J. Parsons, C .S. Penkett and A. J. Shell, *Chem. Rev.,* **1996**, *96*, 195.

- For a recent example of the application of radical deoxygenation for the synthesis of (+)-brazilane see: J. Xu and J. C. Yadan, *Tetrahedron Lett.,* **1996**, *37*, 2421.

75. Total Synthesis of (±)-Myrocin C

M. Y. Chu-Moyer, S. J. Danishefsky and G. K. Schulte, *J. Am. Chem. Soc.*, **1994**, *116*, 11213.

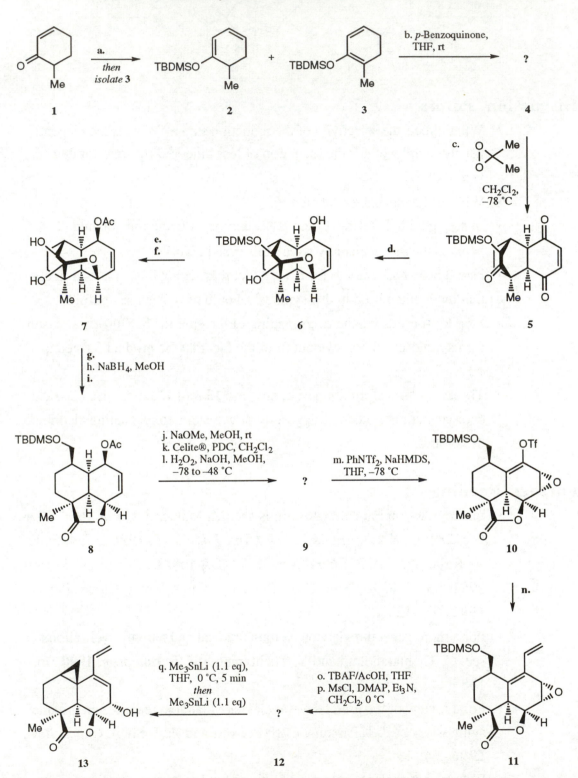

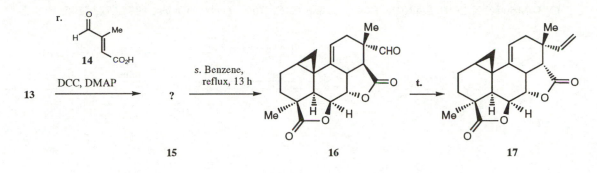

Discussion Points

- What is the mechanism of step **b**?
- Propose an alternative method of carrying out the transformation achieved with 3,3-dimethyldioxirane in step **c**.
- Rationalise the stereoselectivity of the reduction in step **d**.
- Bearing in mind the observation that on treating the tosylate **18** with *tert*-butyl lithium at –78 °C compound **19** was obtained in 74% yield, suggest a mechanism for the formation of **13** from **12**.

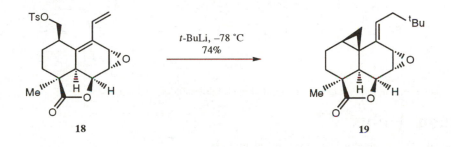

Further Reading

- For an alternative approach to a fragment of myrocin C see: W. Langschwater and H. M. Hoffman, *Liebigs Ann.*, **1995**, *5*, 797.
- For some other recent applications of intramolecular Diels–Alder reactions see: P. J. Ainsworth, D. Craig, A. J. P. White and D. J. Williams, *Tetrahedron* **1996**, *52*, 8937; M. Naruse, S. Aoyagi and C. Kibayashi, *J. Chem. Soc., Perkin Trans., 1*, **1996**, 1113; T.-C. Chou, P.-C. Hong, Y.-F. Wu, W.-Y. Chang, C.-T. Lin and K.-J. Lin, *Tetrahedron*, **1996**, *52*, 6325; C. D. Dzierba, K. S. Zandi, T. Moellera and K. J. Shea, *J. Am. Chem. Soc.*, **1996**, *118*, 4711; M. Lee, I. Ikeda, T. Kawabe, S. Mori and K. Kanematsu, *J. Org, Chem.*, **1996**, *61*, 3406.

76. Synthesis of Staurosporine Aglycon

C. J. Moody, K. F. Rahimtoola, B. Porter and B. C. Ross, *J. Org. Chem.*, **1992**, *57*, 2105.

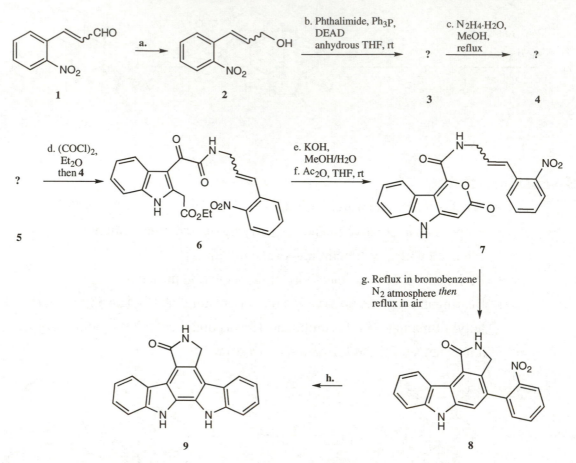

Discussion Points

- What is the mechanism of step **b**?
- Propose a mechanism for the transformation covered in step **f**.
- Simply refluxing compound **7** under an inert atmosphere in bromobenzene gave a product **10** in high yield. In the proton NMR spectrum of this product there were the following signals integrating for single protons; δ 3.28 (dd, $J = 17$ and 7 Hz), δ 3.69 (dd, $J = 17$ and 10 Hz), δ 4.72 (dd, $J = 10$ and 7 Hz). The FAB mass spectrum gave a protonated molecular ion at 346 *m/e*. Suggest a structure for this compound and the motive for a second reflux of the reaction mixture in air.

Further Reading

- For the first total synthesis of staurosporine see: J. T. Link, S. Raghavan and S. J. Danishefsky, *J. Am. Chem. Soc.,* **1995**, *117*, 552.
- A structure–activity study of a series of analogous protein kinase inhibitors was published by J. Zimmermann, T. Meyer and J. W. Lown in *Bioorg. Med. Chem. Lett.*, **1995**, *5*, 497.

77. Synthesis of (20S)-Camptothecin

D. P. Curran, S.-B. Ko and H. Josien, *Angew. Chem., Int. Ed. Engl.*, **1995**, *34*, 2683.

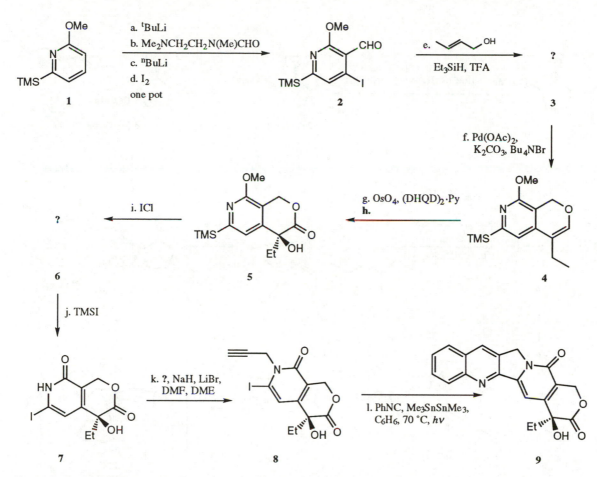

Discussion Points

- Explain the regioselectivity observed in the reaction sequence **a** to **d** and give reasons for the fact that it is possible to use a nucleophilic base such as *n*-BuLi in step **c**.
- What is the mechanism of the reductive etherification step **e**?
- Suggest structures for the intermediates formed in the radical cascade reaction of **8** with phenyl isonitrile.

Further Reading

- For a review on the use of radical reactions in natural product synthesis, see: U. Koert, *Angew. Chem., Int. Ed. Engl.*, **1996**, *35*, 405.
- For a review on the directed *ortho*-metallation reaction, see: V. Snieckus, *Pure Appl. Chem.*, **1990**, *62*, 2047; G. Quéguiner, F. Marsais, V. Snieckus and J. Epsztajn, *Adv. Heterocycl. Chem.*, **1991**, *52*, 187.

78. Synthesis of a Fragment of (+)-Codaphniphylline

C. H. Heathcock, J. C. Kath and R. B. Ruggeri, *J. Org. Chem.*, **1995**, *60*, 1120.

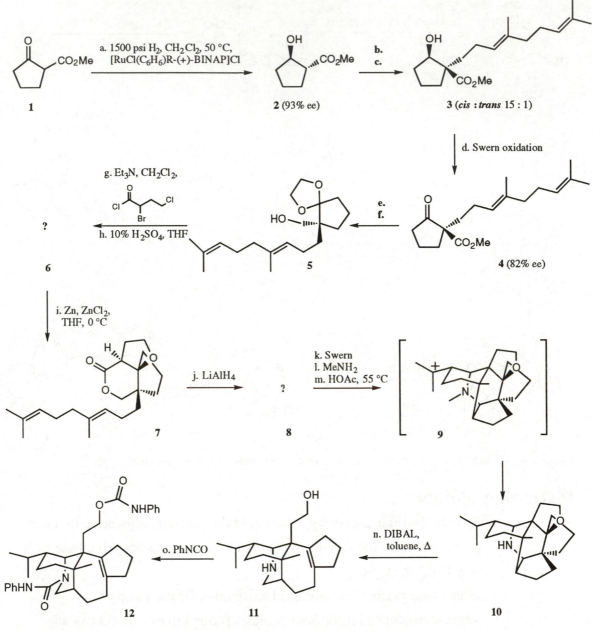

78. Synthesis of a Fragment of (+)-Codaphniphylline

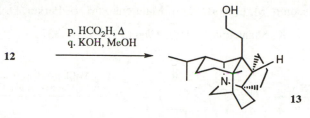

12 $\xrightarrow{\substack{\text{p. HCO}_2\text{H, }\Delta \\ \text{q. KOH, MeOH}}}$ **13**

Discussion Points

- Explain the loss of enantiomeric excess observed in the transformation of **2** into **4**.

- A small amount of compound **14** is formed under the Reformatsky condition of step **i**. Suggest a mechanism for its formation.

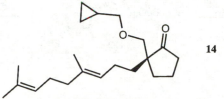

14

- Suggest a mechanism for the formation of compound **10** and for the demethylation of the intermediate carbocationic species **9**.

- What is the role of DIBAL in the Grob fragmentation of step **n**?

- Give a mechanism for the formation of compound **13**. Suggest a reason for the necessity of transforming amine **11** into the urea derivative **12**.

Further Reading

- For a recent review on the Reformatsky reaction, see: A. Fuerstner, *Synthesis*, **1989**, *8*, 571.

- For another application of the Grob fragmentation to the synthesis of azabicyclononanes, see: N. Risch, U. Billerbeck and B. Meyer-Roscher, *Chem. Ber.*, **1993**, *126*, 1137.

- For the X-ray structure analysis of iminium ions formed via Grob fragmentation, see: S. Hollenstein and T. Laube, *Angew. Chem., Int. Ed. Engl.*, **1990**, *102*, 194.

79. The Total Synthesis of Indanomycin

S. D. Burke, A. D. Piscopio, M. E. Kort, M. A. Matulenko, M. H. Parker, D. M. Armistead and

K. Shankaran, *J. Org. Chem.*, **1994**, *59*, 332.

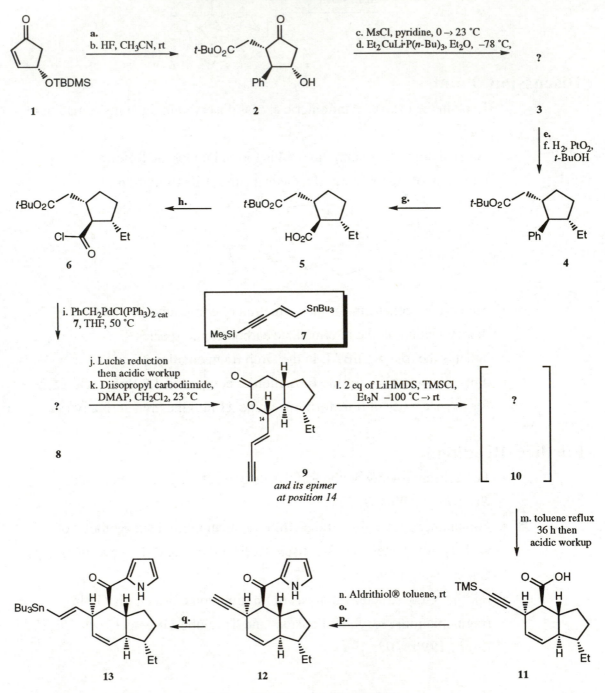

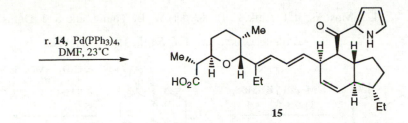

13 — r. **14**, Pd(PPh₃)₄, DMF, 23°C ⟶ **15**

Discussion Points

- Rationalise the stereoselectivity of the formation of compound **2**.
- What is the mechanism of the formation of compound **3** from **2**?
- The Luche reduction of compound **8** was non-selective, resulting, after step **j**, in the formation of separable epimers **9**. How could the undesired epimer be recycled?
- Suggest a mechanism for the formation of compound **11**.
- In a model system **A**, the thermolysis at 105 °C resulted in the formation of an open-chain molecule **B** which on heating at 135 °C was transformed into the desired analogue of bicyclic compound **11**. How does this influence the possible mechanism of formation of compound **11**?

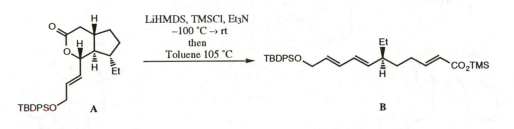

LiHMDS, TMSCl, Et₃N
−100 °C → rt
then
Toluene 105 °C

A ⟶ **B**

- Propose a structure for compound **14**.

Further Reading

- For reviews on Stille and related cross-coupling reactions see; R. Rossi, A. Carpita and F. Bellina, *Org. Prep. Proced. Int.,* **1995**, *27*, 127; T. N. Mitchell, *Synthesis,* **1992**, 803; M. Kosugi and T. Migita, *Trends Org. Chem.,* **1990**, 151; J. K. Stille, *Pure Appl. Chem.,* **1991**, *63*, 419.

80. A Total Synthesis of Taxol

J. J. Masters, J. T. Link, L. B. Snyder, W. B. Young, and S. J. Danishefsky,

Angew. Chem., Int. Ed. Engl., **1995**, *34*, 1723.

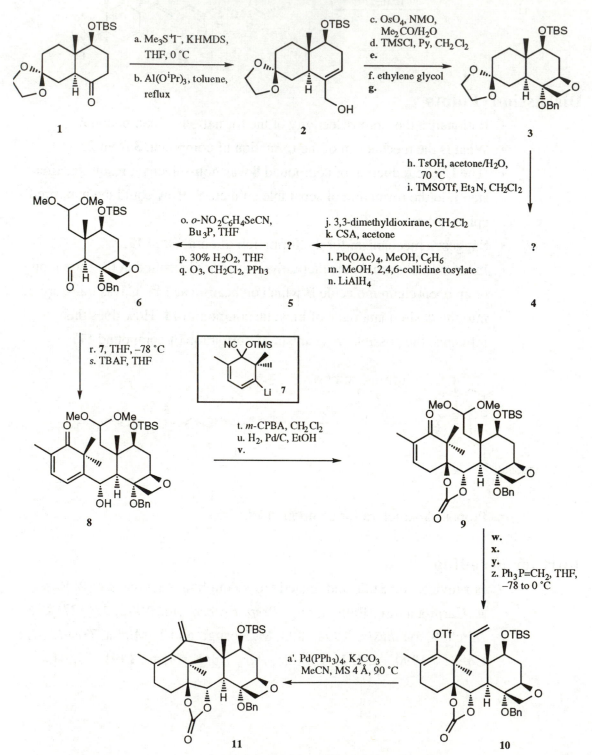

Discussion Points

- What is the mechanism of step **a**?
- What is the purpose of aluminium triisopropoxide in step **b**?
- Give reasons for the selectivity (4:1) obtained in the osmylation step **c**.
- When fluoride ion was used instead of ethylene glycol as the desilylating agent to promote the formation of the oxetane ring in **3**, the main product isolated was **12**. Suggest the mechanism for its formation.

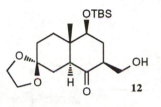

- Explain the diastereoselectivity observed in the formation of compound **8**.
- Explain the stereoselectivity of epoxidation step **t**.
- Give reasons for the regioselectivity observed in the epoxide reduction step **u**.
- Suggest a way of synthesising **7**.

Further Reading

- For a recent review on the Heck reaction, see: W. Cabri and I. Candiani, *Acc. Chem. Res.*, **1995**, *28*, 2.
- For a review on chelation controlled carbonyl addition reactions, see: M. T. Reetz, *Acc. Chem. Res.*, **1993**, *26*, 462.
- For further studies, see: S. J. Danishefsky, J. J. Masters, W. B. Young, J. T. Link, L. B. Snyder, T. V. Magee, D. K. Jung, R. C. A. Isaacs, W. G. Bornmann, *J. Am. Chem. Soc.*, **1996**, *118*, 2843.

81. Total Synthesis of (−)-Grayanotoxin III

T. Kan, S. Hosokawa, S. Nara, M. Oikawa, S. Ito, F. Matsuda and H. Shirahama

J. Org. Chem., **1994**, *59*, 5532.

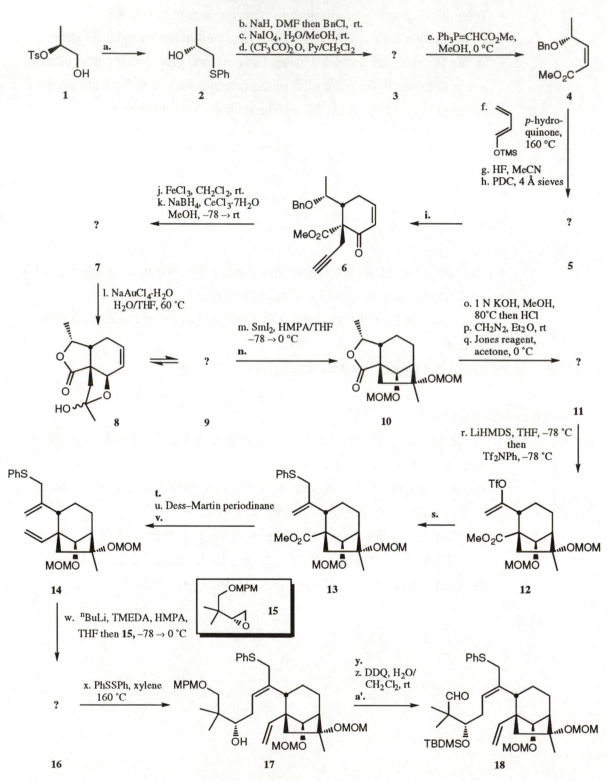

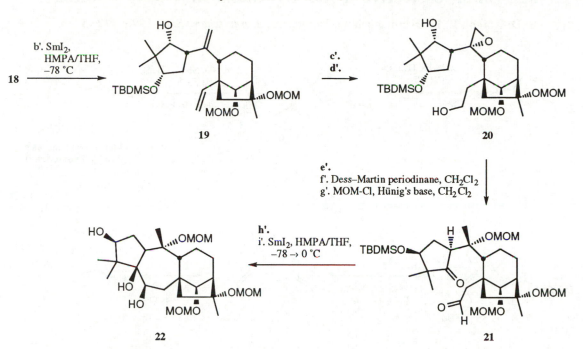

Discussion Points

- Propose a mechanism for the sequence carried out in steps **c** and **d**.
- The epoxidation of compound **19** is highly stereoselective. What is likely to be the major factor in controlling this process?
- Give an alternative method of generating a 1,2-diol from two carbonyl groups as carried out in step **i´**.

Further Reading

- For reviews on applications of samarium iodide to organic synthesis see: G. A. Molander and C. R. Harris, *Chem. Rev.*, **1996**, *96*, 307; G. A. Molander, *Org. React.*, **1994**, *46*, 211;
- For further applications of samarium iodide promoted ketone–olefin couplings see: M. Kawatsura, K. Hosaka, F. Matsuda and H. Shirahama, *Synlett*, **1995**, 729.
- For a recent review on synthetic transformations of vinyl and aryl triflates see: K. Ritter, *Synthesis*, **1993**, 735.
- For an approach to a related system involving a sequential Tebbe olefination and Claisen rearrangement as a key transformation see: S. Borrelly and L. A. Paquette, *J. Am. Chem. Soc.*, **1996**, *118*, 727.
- For a review on the hydration of acetylenes without using mercury (step **k**) see: I. K. Meier and J. A. Marsella, *J. Mol. Catal.*, **1993**, *78*, 31.
- For a further application of $FeCl_3$ promoted debenzylation–lactonisation see: R. Zemribo, M. S. Champs and D. Romo, *Synlett*, **1996**, 278.

82. Enantioselective Total Synthesis of (−)-Strychnine

S. D. Knight, L. E. Overman and G. Pairaudeau, *J. Am. Chem. Soc.*, **1995**, *117*, 5776.

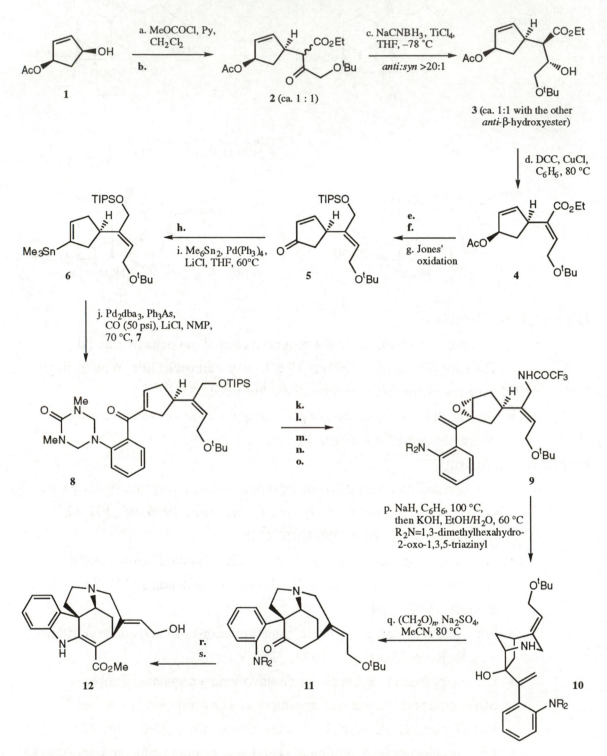

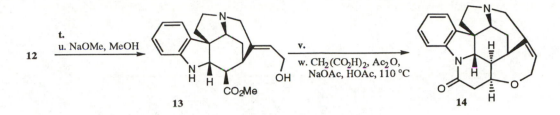

Discussion Points

- Suggest a structure for compound **7**.
- Give a method for the preparation of starting material **1** or its enantiomer.
- Explain the diastereoselectivity in the reduction of **2** to **3**.
- What is the mechanism for the elimination step **d**?
- What is the mechanism for step **q**?
- What is the purpose of sodium methoxide in step **u**?

Further Reading

- For an analysis of the published total syntheses of strychnine, see: U. Beifuss, *Angew. Chem., Int. Ed. Engl.*, **1994**, *33*, 1144.
- For a study on stereoselective BCl₃ and TiCl₄-mediated reductions of β-hydroxyketones, see: C. R. Sarko, S. E. Collibee, A. L. Knorr and M. DiMare, *J. Org. Chem.*, **1996**, *61*, 868.
- For a brief review on the aza-Cope–Mannich reaction sequence, see: L. Overman, *Acc. Chem. Res.*, **1992**, *25*, 352.
- For the use of organotin reagents in the synthesis of carbonyl compounds, see: M. Kosugi and T. Toshihiko, *Trends Org. Chem.*, **1990**, *1*, 151.